THROUGH TWO DOORS AT ONCE

ALSO BY ANIL ANANTHASWAMY

The Edge of Physics

The Man Who Wasn't There

THROUGH TWO DOORS AT ONCE

The Elegant Experiment That Captures
the Enigma of Our Quantum Reality

Anil Ananthaswamy

DUTTON

DUTTON

An imprint of Penguin Random House LLC
375 Hudson Street
New York, New York 10014

Copyright © 2018 by Anil Ananthaswamy

Portions of chapters 5 and 6 appeared in *New Scientist* magazine. Bohmian trajectories in chapter 6 reproduced with permission from Chris Dewdney. The de Broglie-Bohm and the many-interacting worlds trajectories in the epilogue reproduced with permission granted by Howard Wiseman on behalf of his coauthors.

Illustrations credit: Roshan Shakeel

LIBRARY OF CONGRESS CATALOGING-IN-PUBLICATION DATA
Names: Ananthaswamy, Anil, author.
Title: Through two doors at once : the elegant experiment that captures the enigma
    of our quantum reality / Anil Ananthaswamy.
Description: New York, New York : Dutton, an imprint of Penguin Random House LLC,
    [2018] | Includes bibliographical references and index.
Identifiers: LCCN 2018008272 | ISBN 9781101986097 (hardcover) | ISBN 9781101986110
    (ebook) | Subjects: LCSH: Quantum theory—Popular works. | Wave theory of light—
    Popular works. | Reality—Popular works.
Classification: LCC QC174.123.A53 2018 | DDC 530.12—dc23
LC record available https://lccn.loc.gov/2018008272

Printed in the United States of America
10 9 8 7 6 5 4 3 2 1

Set in Warnock Pro with HK Grotesk
Designed by Daniel Lagin

To my parents

Allow me to express now, once and for all, my deep respect for the work of the experimenter and for his fight to wring significant facts from an inflexible Nature . . . [which] says so distinctly "No" and so indistinctly "Yes" to our theories

—Hermann Weyl, German mathematician, 1885–1955

CONTENTS

CONTENTS

THROUGH TWO DOORS AT ONCE

Prologue

THE STORY OF NATURE TAUNTING US

The office is simply the most uncluttered of any physicist's office I have ever seen. There's a chair alongside a small table, with nothing on it. No books, no papers, no lamp, no computer, nothing. A sofa graces the office. Large windows overlook a small lake, the trees around which are bare, except for a few stragglers that are holding on to their fall foliage, defying the approaching winter in this part of Ontario, Canada. Lucien Hardy puts his laptop on the table—pointing out that he does most of his work in cafés and figures that all he needs in his office is a café-like small table to set down his laptop.

There is the obligatory blackboard, taking up most of one wall of his office. It doesn't take long for Hardy to spring up and start chalking it up with diagrams and equations—something that most of the quantum physicists I meet seem inclined to do.

We start talking about some esoteric aspect of quantum physics, when he stops and says, "I started off the wrong way." To reset our discussion, he says, "Imagine you have a factory and they make bombs." He has my attention.

He writes two names on the blackboard: Elitzur and Vaidman. He is talking about something called the Elitzur-Vaidman bomb puzzle. Named after two Israeli physicists, the puzzle exemplifies the counterintuitive nature of the quantum world in ways that non-physicists can appreciate. It confounds physicists too in no small measure.

The problem goes something like this. There's a factory that makes bombs equipped with triggers. The triggers are so sensitive that a single particle, any particle, even a particle of light, can set them off. There's a big dilemma, however. The factory's assembly line is faulty. It's churning out both good bombs with triggers and bad bombs without triggers. Hardy writes them as "good" and "bad" and quips about the quotation marks: "Obviously, you may have a different moral perspective on it."

The task is to identify the good bombs. This means having to check whether the bombs have triggers. But examining each bomb isn't the correct strategy, because in order to do so, you'd need to shine light on it, however faint, and that would cause a good bomb to explode. The only ones left unexploded would be the duds without triggers.

So, how does one solve this problem? If it helps, we are allowed one concession: we can detonate some bombs, as long as we are left with some good, undetonated bombs.

From our everyday experience of how the world works, this is an impossible problem to solve. But the quantum world—the world of very small things like molecules and atoms and electrons and protons and photons—behaves in bizarre ways. The physics that governs the behavior of this microscopic world is called quantum physics or quantum mechanics. And we can use quantum physics to find good bombs without setting them off. Even with a simple setup, it's possible to salvage about half the good bombs. It involves using a modern variation of a 200-year-old experiment.

Called the double-slit experiment, it was first done in the early 1800s to challenge Isaac Newton's ideas about the nature of light. The experiment took center stage again in the early twentieth century, when two of the founders of quantum physics, Albert Einstein and Niels Bohr, grappled with its revelations about the nature of reality. In the 1960s, Richard Feynman extolled its virtues, saying that the double-slit experiment contained all of the mysteries of the quantum world. A simpler and more elegant experiment would be hard to find, the workings of which a high school student can grasp, yet profound enough in its implications to bewilder brains like Einstein's and Bohr's, a confusion that continues to this day.

This is the story of quantum mechanics from the perspective of one classic experiment and its subtle, sophisticated variations (including one that, as we'll see, solves the Elitzur-Vaidman bomb puzzle), whether these variations are carried out as thought experiments by luminous minds or painstakingly performed in the basement labs of physics departments. It's the story of nature taunting us: catch me if you can.

1

THE CASE OF
THE EXPERIMENT
WITH TWO HOLES

Richard Feynman Explains
the Central Mystery

There is nothing more surreal, nothing more abstract than
reality.

—**Giorgio Morandi**

Richard Feynman was still a year away from winning his Nobel
Prize. And two decades away from publishing an endearing
autobiographical book that introduced him to non-physicists as a
straight-talking scientist interested in everything from cracking
safes to playing drums. But in November 1964, to students at Cornell University in Ithaca, New York, he was already a star and they
received him as such. Feynman came to deliver a series of lectures.
Strains of "Far above Cayuga's Waters" rang out from the Cornell
Chimes. The provost introduced Feynman as an instructor and
physicist par excellence, but also, of course, as an accomplished
bongo drummer. Feynman strode onto the stage to the kind of
applause reserved for performing artists, and opened his lecture

with this observation: "It's odd, but in the infrequent occasions when I have been called upon in a formal place to play the bongo drums, the introducer never seems to find it necessary to mention that I also do theoretical physics."

By his sixth lecture, Feynman dispensed with any preamble, even a token "Hello" to the clapping students, and jumped straight into how our intuition, which is suited to dealing with everyday things that we can see and hear and touch, fails when it comes to understanding nature at very small scales.

And often, he said, it's experiments that challenge our intuitive view of the world. "Then we see unexpected things," said Feynman. "We see things that are very far from what we could have imagined. And so our imagination is stretched to the utmost—not, as in fiction, to imagine things which aren't really there. But our imagination is stretched to the utmost just to comprehend those things which are there. And it's this kind of a situation that I want to talk about."

The lecture was about quantum mechanics, the physics of the very small things; in particular, it was about the nature of light and subatomic bits of matter such as electrons. In other words, it was about the nature of reality. Do light and electrons show wavelike behavior (like water does)? Or do they act like particles (like grains of sand do)? Turns out that saying yes or no would be both correct and incorrect. Any attempt to visualize the behavior of the microscopic, subatomic entities makes a mockery of our intuition.

"They behave in their own inimitable way," said Feynman. "Which, technically, could be called the 'quantum-mechanical' way. They behave in a way that is like nothing that you have ever seen before. Your experience with things that you have seen before is inadequate—is incomplete. The behavior of things on a very tiny scale is simply different. They do not behave *just* like particles. They do not behave *just* like waves."

But at least light and electrons behave in "exactly the same" way, said Feynman. "That is, they're both screwy."

Feynman cautioned the audience that the lecture was going to be difficult because it would challenge their widely held views about how nature works: "But the difficulty, really, is psychological and exists in the perpetual torment that results from your saying to yourself 'But how can it be like that?' Which really is a reflection of an uncontrolled, but I say utterly vain, desire to see it in terms of some analogy with something familiar. I will not describe it in terms of an analogy with something familiar. I'll simply describe it."

And so, to make his point over the course of an hour of spell-binding oratory, Feynman focused on the "one experiment which has been designed to contain all of the mystery of quantum mechanics, to put you up against the paradoxes and mysteries and peculiarities of nature."

It was the double-slit experiment. It's difficult to imagine a simpler experiment—or, as we'll discover over the course of this book, one more confounding. We start with a source of light. Place in front of the source a sheet of opaque material with two narrow, closely spaced slits or openings. This creates two paths for the light to go through. On the other side of the opaque sheet is a screen. What would you expect to see on the screen?

The answer, at least in the context of the world we are familiar with, depends on what one thinks is the nature of light. In the late

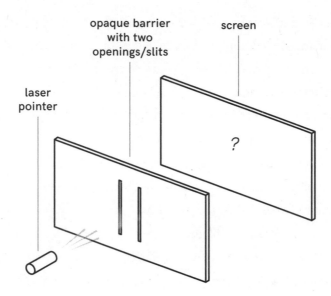

seventeenth century and all of the eighteenth century, Isaac Newton's ideas dominated our view of light. He argued that light was made of tiny particles, or "corpuscles," as he called them. Newton's "corpuscular theory of light" was partly formulated to explain why light, unlike sound, cannot bend around corners. Light must be made of particles, Newton argued, since particles don't curve or bend in the absence of external forces.

In his lecture, when Feynman analyzed the double-slit experiment, he first considered the case of a source firing particles at the two slits. To accentuate the particle nature of the source, he urged the audience to imagine that instead of subatomic particles (of which electrons and particles of light would be examples), we were to fire bullets from a gun—which "come in lumps." To avoid too much violent imagery (what with bombs in the prologue, and a thought experiment with gunpowder to come), let's imagine a source that spews particles of sand rather than bullets; we know that sand comes in lumps, though the lumps are much, much smaller than bullets.

First, let's do the experiment with either the left slit or the right slit closed. Let's take it that the source is firing grains of sand at high enough speeds that they have straight trajectories. When we do this, the grains of sand that get through the slits mostly hit the region of the screen directly behind the open slit, with the numbers tapering off on either side. The higher the height of the graph, the more the number of grains of sand reaching that location on the screen.

Now, what should we see if both slits are open? As expected, each grain of sand passes through one or the other opening and reaches the other side. The distribution of the grains of sand on the far screen is simply the sum of what goes through each slit. It's a

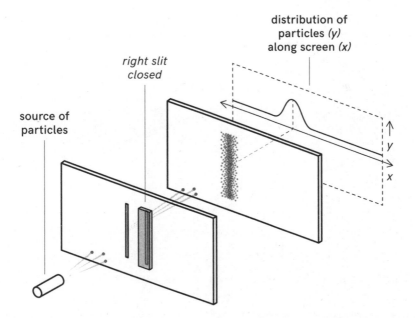

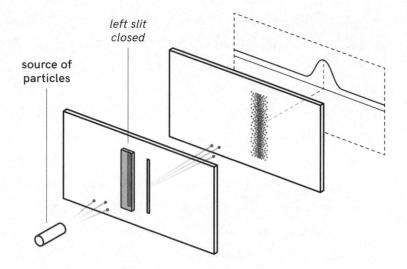

demonstration of the intuitive and sensible behavior of the non-quantum world of everyday experience, the classical world described so well by Newton's laws of motion.

To be convinced that this is indeed what happens with particles

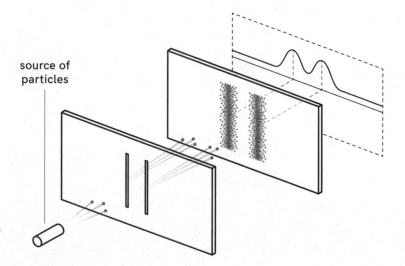

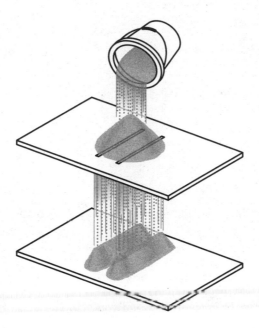

of sand, let's orient the device such that the sand is now falling down onto the barrier with two slits. Our intuition clearly tells us that two mounds should form beneath the two openings.

Turning the experiment back to its original position, let's dispense with the sand and consider a source that's emitting light, and assume that light's made of Newtonian corpuscles. Informed by our experiment with sand particles, we'd expect to see two strips of light on the screen, one behind the right slit and one behind the left slit, each strip of light fading off to the sides, leading to a distribution of light that is simply the sum of the light you'd get passing through each slit.

Well, that's not what happens. Light, it seems, does not behave as if it's made of particles.

Even before Newton's time, there were observations that

challenged his theory of the particle nature of light. For example, light changes course when going from one medium to another—say, from air to glass and back into air (this phenomenon, called refraction, is what allows us to make optical lenses). Refraction can't be easily explained if you think of light as particles traveling through air and glass, because it requires positing an external force to change the direction of light when it goes from air to glass and from glass to air. But refraction can be explained if light is thought of as a wave (the speed of the wave would be different in air than in glass, explaining the change in direction as light goes from one type of material to another). This is exactly what Dutch scientist Christiaan Huygens proposed in the 1600s. Huygens argued that light is a wave much like a sound wave, and since sound waves are essentially vibrations of the medium in which they are traveling, Huygens argued that light too is made of vibrations of a medium called ether that pervades the space around us.

This was a serious theory put forth by an enormously gifted scientist. Huygens was a physicist, astronomer, and mathematician. He made telescopes by grinding lenses himself, and discovered Saturn's moon Titan (the first probe to land on Titan, in 2005, was named Huygens in his honor). He independently discovered the Orion nebula. In 1690, he published his *Traité de la Lumière* (*Treatise on Light*), in which he expounded his wave theory of light.

Newton and Huygens were contemporaries, but Newton's star shone brighter. After all, he had come up with the laws of motion and the universal law of gravitation, which explained everything from the arc of a ball thrown across a field to the movement of planets around the sun. Besides, Newton was a polymath of considerable renown (as a mathematician, he gave us calculus, and even ventured into chemistry, theology, and writing biblical commentaries, not to mention all his work in physics). It was no wonder that his corpuscular theory of light, despite its shortcomings, overshadowed Huygens's ideas of light being wavelike. It'd take another polymath to show up Newton when it came to understanding light.

Thomas Young has been called "The Last Man Who Knew Everything." In 1793, barely twenty years of age, he explained how our eyes focus upon objects at different distances, based partly on his own dissection of an ox's eyes. A year later, on the strength of that work, Young was made a Fellow of the Royal Society, and in 1796 he became "doctor of physic, surgery, and midwifery." When he was in his forties, Young helped Egyptologists decipher the Rosetta stone (which had inscriptions in three scripts: Greek, hieroglyphics, and something unknown). And in between becoming a medical doctor, getting steeped in Egyptology, and even studying Indo-European languages, Young delivered one of the most intriguing lectures in the history of physics. The venue was the Royal Society of London, and the date, November 24, 1803. Young stood in front of that august audience, this time as a physicist describing a simple and elegant homespun experiment, which, in his mind, had unambiguously established the true nature of light and proved Newton wrong.

"The experiments I am about to relate . . . may be repeated with great ease, whenever the sun shines," Young told the audience.

Whenever the sun shines. Young wasn't overstating the simplicity of his experiment. "I made a small hole in a window-shutter, and covered it with a piece of thick paper, which I perforated with a fine needle," he said. The pinhole let through a ray of light, a sunbeam. "I brought into the sunbeam a slip of card, about one-thirtieth of an inch in breadth, and observed its shadow, either on the wall, or on other cards held at different distances."

If light is made of particles, Young's "slip of card" would have cast a sharp shadow on the wall in front, because the card would have blocked some of the particles. And if so, Newton would have been proved right.

If, however, light is made of waves, as Huygens claimed, then the card would have merely impeded the waves, like a rock impedes flowing water, and the wave would have gone around the card, taking two paths, one on either side of the card. The two paths of light would eventually recombine at the wall opposite the window shutter to create a characteristic pattern: a row of alternating bright and dark stripes. Such stripes, also known as interference fringes, are created when two waves overlap. Crucially, the central fringe would be bright, exactly where you'd expect a dark shadow if light were made of particles.

We know about interference from our everyday experience of waves of water. Think of an ocean wave hitting two openings in a coastal breakwall. New waves emerge from each opening (a process called diffraction) and travel onward, where they overlap and interfere with each other. In regions where the crests of both waves arrive at the same time, there's constructive interference and the water is at its highest (analogous to bright fringes of light); and in regions where the crest from one wave arrives at the same time as the trough of the other, the waves cancel each other out and there's destructive interference (corresponding to dark fringes).

Young saw such optical interference fringes. Specifically, since he was working with sunlight, which contains light of all colors, he saw a central region that was flanked by fringes of colors. The central region, upon closer inspection, was seen to be made of light and dark fringes. The numbers of these fringes and their widths depended on how far away the pinhole in the window shutter was from the screen or wall. And the middle of the central region was always white (a bright fringe). He had shown that light is wavelike.

There must have been disbelief in the audience, for Young was going against Newton's ideas. Even before Young's lecture, articles written anonymously in the *Edinburgh Review* had been heavily critical of his work. The author, who turned out to be a barrister named Henry Brougham (he became Lord Chancellor of England in 1830), was scathing, calling Young's work "destitute of every species of merit" and "the unmanly and unfruitful pleasure of a boyish and prurient imagination."

It was anything but. Soon enough, Young's ideas got further support from other physicists. His experiment led to what's now called the double-slit experiment and was in fact the first formulation of it—the very same experiment whose virtues Feynman extolled during his lecture at Cornell. In the more standard double-slit experiment, Young's sunbeam is replaced by a source of light. And instead of a "slip of card" placed in the sunbeam's path to create two paths for the light, the double-slit experiment creates two paths of light by letting the light fall on an opaque barrier with two narrow slits or openings through which the light can pass. And on the screen on the far side, you see an interference pattern, essentially fringes similar to what Young saw on the wall opposite the window

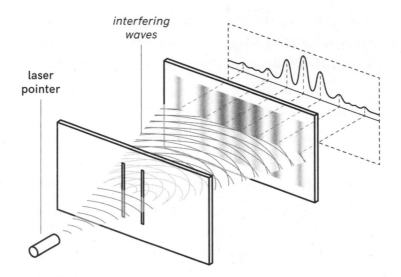

interfering
waves

laser
pointer

shutter (if the screen is a photographic plate, or a piece of glass coated with photosensitive material, then the image can be thought of as a film negative: dark regions will form where the film is being exposed to light). You don't see just two strips tapering away, which you'd expect if light behaved as if it came in lumps. It's behaving like a wave.

So, well before quantum physics was even a gleam in anyone's eyes, Young had seemingly settled the debate between Newton and Huygens (despite skeptics who continued to favor Newton). Young came down in favor of Huygens's light-is-a-wave idea. And so things stood until the quantum revolution.

The revolution began with bewildering discoveries in the early 1900s, including Albert Einstein's 1905 assertion that light should be thought of as being made of particles, because it was the only way to explain a phenomenon known as the photoelectric effect (which helps us convert sunlight into electricity, giving us the technology of solar panels). These particles of light came to be called photons. For any given frequency or color of light, a photon of light is the smallest unit of energy, and it cannot be divided any further: the light cannot come with any less energy than contained in one photon. Einstein's argument is somewhat involved, but for now, if we accept the idea that there are certain situations in physics where you have to treat light as made of particles, then the double-slit experiment starts challenging our intuitive sense of reality.

Feynman spoke of the double-slit experiment as embodying the "central mystery" of quantum mechanics. To show why, he replaced the gun shooting bullets (or, in our case, grains of sand) with a source of electrons. Everyone in the 1960s knew that electrons came in lumps. They are one of the many types of elementary particles that make up the subatomic world, including photons. We'll use photons instead of electrons. The fact that the experiment, its results, and its implications don't change whether we are using photons, which are particles of light without any mass, or electrons, which are particles of matter with some mass, leads to its own set of mystifying questions. As Feynman said, both are screwy in the same way.

Here's what happens if you use photons. Unlike what we got with particles of sand, you don't get two bands of light on the screen. Instead, you get fringes, similar to the interference pattern that Young observed, suggesting that photons are behaving like waves. To get a sharply defined set of fringes, it's best to use light of one color. So the source can be, say, streaming out an intense beam of photons of red light that pass through the double slit.

distribution of electrons or
photons has peaks and valleys.
The peaks are places where
more particles land, and the
valleys are places where fewer
or no particles land

source of
particles

When both slits are open, you get the interference pattern, suggesting that light (which we know now is made of particles) is going
through both slits. But if you close one of the two slits (doesn't matter which one), the interference pattern disappears, clearly suggesting that light is going through only one slit and there's nothing for
it to interfere with.

The experiment, however, really starts messing with our minds
when we consider a source that emits one photon at a time. We'll
come to the ways in which physicists invented sources to do that. It
wasn't possible in 1964, when Feynman was giving his lecture. For
now, let's assume we have such a source in hand. If so, each photon

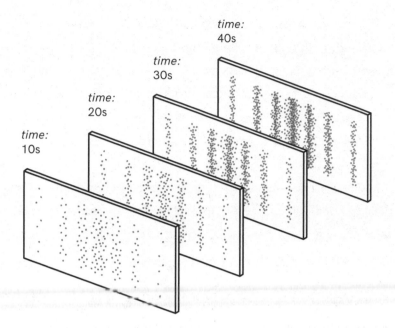

goes through the apparatus, and we make sure there's only one photon passing through the setup at a time. The photon hits the photographic plate on the far side and creates a spot. If we let enough spots accumulate, our intuition says that these photons should act like grains of sand and line up behind each slit. There should be no interference pattern.

We'd be wrong. As it happens, even though each photon seems to be landing at some random position, fringes emerge when enough photons have made their mark on the photographic plate. Each photon makes a dark spot on the plate; places where the photons mostly land become dark stripes, and fringes build up over time.

This is somewhat curious. It's clear that we can get an interference pattern when one wave interferes with another. But our photons are going through the apparatus one by one. There's no interference between one photon and the next, or the first photon and the tenth, and so on. Each photon is on its own. Nonetheless, each photon is mostly landing on the photographic plate at those positions that eventually become regions of constructive interference and mostly avoiding those places that become regions of destructive interference. We get interference fringes. It's *as if* each photon is exhibiting wavelike behavior, *as if* it's interfering with itself.

This is happening even though we create each photon as a particle, and detect it on a photographic plate as a particle: the results *seem to* suggest that between the creation and detection, each particle acts like a wave, and somehow goes through both slits simultaneously. How else do you explain the interference pattern?

If that's not mysterious enough, consider what happens if we try to find out which slit a photon goes through (our intuition, after all, says that it surely went through just one slit, not both). Say you have a mechanism for detecting the passage of a photon through one or the other slit without destroying the photon. If you do that, the interference pattern goes away (meaning the photon stops behaving like a wave and starts acting like a particle)—and you get a pattern that's simply the sum of the "lumps" going through each slit. Stop trying to sneak a peek at the photon's path and it goes back to behaving like a wave—the interference pattern reemerges.

There's yet another way to appreciate this mystery. When you are not looking at the photons' paths, individual photons almost never go to certain places on the photographic plate—the places that eventually become regions of destructive interference. But if you start monitoring their paths, they will go to the very locations that they otherwise shun. What's going on?

The curious behavior continues. If you were to fire grains of sand at the double slit, and if you knew everything about the initial conditions of each grain of sand (its initial velocity, the angle at which it leaves the sand gun, etc.), you can predict using Newton's laws exactly where the grain of sand will end up on the screen opposite the double slit, taking into account any deflections due to the interaction with the slits. This is how physics is supposed to work. But you can't do that with photons (or electrons, or anything quantum mechanical for that matter).

Even if you have all the information about a single photon as it leaves the source and goes toward the double slit, you can only calculate the probability of the photon landing on a certain part of the photographic plate. For example, the photon could land at any one of the many regions of constructive interference—but there's no way to tell exactly where any particular photon will go. Nature, at its deepest, seems inherently nondeterministic. Or is it merely hiding its secrets, and we haven't dug deep enough yet?

The questions pile up. Between the production of the photon and its eventual detection, both proofs of its particle nature, the photon ostensibly behaves like a wave if we choose not to look at which path it takes, and as a particle otherwise. Does the photon "know" we are looking at its wave nature or particle nature? If so, how? And can we fool the photon, say, by not revealing our hand until it has crossed the double slit as a wave, and only then choosing to see which slit it went through, thus examining its particle-like behavior?

Maybe there is a simpler answer: that the photon is always a particle and always goes through one or the other slit. And something else, something that our standard theories don't account for, goes through both slits to produce the wavelike behavior. In that case, what is that something?

If it crossed your mind that human consciousness is somehow involved in causing the photon to behave one way or the other, you wouldn't be alone in thinking so. As often happens when confronted with two mysteries (in this case the odd behavior of the quantum world and the inexplicable nature of consciousness), it's almost human nature to want to conflate the two.

It'd be twenty years on from Feynman's lecture at Cornell that the double-slit experiment would be done using single photons. It was an example of how, from Young's efforts in the early 1800s to modern versions, physicists continue to use the double-slit experiment to understand the nature of reality. The experiment hasn't changed in its conceptual simplicity for more than two hundred years, but it has become technologically more and more sophisticated, as experimenters keep thinking of clever ways to trick nature into revealing its profoundest secrets.

2

WHAT DOES IT MEAN
"TO BE"?

The Road to Reality,
from Copenhagen to Brussels

The idea of an objective real world whose smallest parts exist objectively in the same sense as stones or trees exist, independently of whether or not we observe them . . . is impossible.

—**Werner Heisenberg**

Q uantum physics has been with us for about a century. But for almost two centuries before the birth of quantum physics, our ideas of how nature works were governed by laws discovered by Isaac Newton. He elucidated his laws in the *Principia*, an astonishing treatise published in 1687. Crucial to the Newtonian conception of nature was that it was made of particles of matter whose dynamics were governed by the forces acting upon them, including the mutually attractive force of gravity. Light too was regarded as having particle nature, though this was debated. Huygens, Young, and others challenged this, arguing for light's wave nature. So, while the Newtonian universe was one of particles of matter, light stood apart, its place in the categories of things that make up the world—the ontology of the world—somewhat unclear.

A French prince and physicist, Louis de Broglie, centuries later, would recount this time in the history of physics rather eloquently: "When Light reaches us from the sun or the stars it comes to the eye after a journey across vast spaces void of Matter. It follows from this that Light can cross empty space without difficulty . . . it is not bound up with any motion of Matter. Hence a description of the physical world would remain incomplete unless we were to add to Matter another reality independent of it. This entity is Light. Now what is Light? What is its structure?"

As de Broglie wrote, such questions were looming large in the 1860s, when Scottish scientist James Clerk Maxwell developed the mathematical foundation for physicists to start thinking of light as a wave.

Maxwell's work first involved unifying electricity and magnetism, which until then had been viewed as separate forces, into one force. Building on earlier work by the English physicist and chemist Michael Faraday, Maxwell came up with a theory combining electricity and magnetism, and predicted that they move as one electromagnetic wave. He presented these ideas on December 8, 1864, to the Royal Society of London. The ontology of nature had changed. In addition to particles, it now included electromagnetic fields—oscillations of energy—that moved at the speed of light. Particles were localized, but fields were diffuse and could spread and exert an influence far, far from where they originated.

Maxwell argued that light too is an electromagnetic wave. But his ideas met with some resistance. While physicists could imagine electromagnetic waves moving through a medium, such as a wire, they had trouble envisaging light as an electromagnetic wave moving through the vacuum of space, as it would have to.

But even before questions about the nature of light could be answered, Maxwell's hypothesis about electromagnetism had to be proved. In 1879, the Prussian Academy of Sciences (in Berlin) put out a call for what came to be called the Berlin Prize problem. The prize was for experimentally verifying Maxwell's ideas. Entries were due by March 1, 1882, with the winner to be awarded 100 ducats (a ducat was either a gold or a silver coin used in Europe during the Middle Ages, and even into the nineteenth and early twentieth centuries). One of the scientists thought most likely to win the prize was the prodigiously talented German physicist Heinrich Hertz. That year, Hertz considered the problem but gave up on it, for he could see no clear experimental way forward. "But in spite of having abandoned the solution at that time, I still felt ambitious to discover it by some other method," he later wrote.

No one won the prize in 1882.

Hertz, however, in just a few years solved the puzzle. He designed an experiment that proved Maxwell correct. The experiment involved building a transmitter of electromagnetic waves, and a receiver—and showing that these invisible waves did indeed exist and could propagate through air. Hertz had inadvertently discovered radio waves.

When asked about the usefulness of such waves, Hertz reportedly said, "It is of no use whatsoever. This is just an experiment that proves Maestro Maxwell was right. We just have these mysterious electromagnetic waves that we cannot see with the naked eye. But they are there."

Hertz's experiments validated Maxwell's theory of electromagnetism. Eventually, it would become clear that light too is an electromagnetic wave. It consists of an electric field and a magnetic field, which each vibrate in mutually perpendicular planes. And light itself travels in a direction that is perpendicular to both the constituent electric and magnetic fields. The frequency of vibration, or the frequency of the electromagnetic wave (v), turns out to be equal to the velocity of light (c) divided by its wavelength (λ).

But while doing this experiment, Hertz stumbled upon another curious phenomenon that would, within a decade, challenge the light-is-a-wave argument. The phenomenon is now called the photoelectric effect. When light falls on certain metals, it can eject electrons. Most important, for a given metal, the electrons are ejected

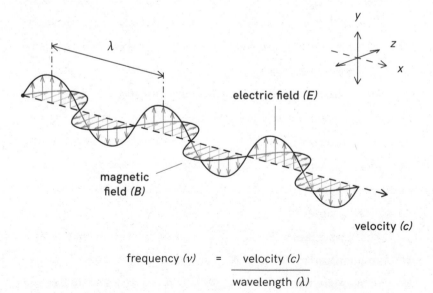

$$\text{frequency } (v) \quad = \quad \frac{\text{velocity } (c)}{\text{wavelength } (\lambda)}$$

only when the light is above a threshold frequency unique to that metal. Below that frequency, regardless of how much light falls on the metal, no electrons are ejected. Above the threshold frequency, two things happen. One is that the number of electrons ejected increases as the intensity of the incident light increases. The other is that increasing the frequency of the light increases the energy of the ejected electrons.

Hertz, however, had seen only glimmers of this phenomenon. His receiver, which was intercepting invisible radio waves, worked better when it was illuminated by light, compared to when it was in darkness inside an enclosure. The radio waves had nothing to do with the light, yet something about the light was influencing the receiver. In a letter he wrote to his father in July 1887, Hertz was characteristically modest about his finding: "To be sure, it is a discovery, because it deals with a completely new and very puzzling phenomenon. I am of course less capable of judging whether it is a beautiful discovery, but of course it does please me to hear others call it that; it seems to be that only the future can tell whether it is important or unimportant."

It's not surprising that what Hertz had observed could not be explained at the time. Physicists were yet to discover electrons, let alone understand the photoelectric effect in all its intricacies. Even as late as the early 1890s, our conception of reality was that atoms were the smallest constituents of the material world, but the structure of the atom was still unknown. The discovery of the electron and other important milestones lay on the path from Hertz to Einstein to quantum mechanics.

Hertz, sadly, didn't live to see any of them. He died on January

1, 1894. An obituary in the journal *Nature* recounted his last days: "A chronic, and painful, disease of the nose spread . . . and gradually led to blood poisoning. He was conscious to the last, and must have been aware that recovery was hopeless; but he bore his sufferings with the greatest patience and fortitude." Hertz was only thirty-seven. His mentor, Hermann von Helmholtz (who would himself die later that year) wrote in the preface to Hertz's monograph *The Principles of Mechanics:* "Heinrich Hertz seemed to be predestined to open up to mankind many of the secrets that nature had hitherto concealed from us; but all these hopes were frustrated by the malignant disease which . . . robbed us of this precious life and of the achievements which it promised."

The secrets of nature that Hertz would surely have helped discover came thick and fast. The first one was the discovery of the electron, thanks to something called a cathode ray tube. The tube—essentially a sealed glass cylinder with electrodes on either end, and from which much of the air had been removed—was a scientific curiosity in the mid-nineteenth century. When a high voltage was applied across the electrodes, the tube would light up, and scientists reveled in showing these off to lay audiences. Soon, physicists discovered that pumping out more air, but not all of it, revealed something dramatic: rays seemed to emerge from the negative electrode (the cathode) and streak across to the positive electrode (the anode).

Three years after Hertz's death, the English physicist J. J. Thomson, using a series of elegant experiments, showed unequivocally that these rays were constituents of matter that were smaller than atoms, and their trajectories could be bent by an electric field in ways that proved the rays had negative charge. Thomson had discovered the electron. He, however, called them corpuscles. Thomson speculated these were literally bits of atoms. Not everyone agreed with his pronouncements. "At first there were very few who believed in the existence of these bodies smaller than atoms," he would later say. "I was even told long afterwards by a distinguished physicist who had been present at my lecture at the Royal Institution that he thought I had been 'pulling their legs.'"

Such doubts aside, Thomson changed our conception of the atom forever.

Meanwhile, after Hertz had made his initial discovery of the photoelectric effect, his assistant, Philipp Lenard, took up the cause. He was a fantastic experimentalist. His experiments clearly showed that ultraviolet light falling on metals produced the same kind of particles as seen in the cathode ray tubes: electrons. Crucially, the velocity of these electrons (and hence their energy) did not depend on the intensity of the incident light. Lenard, however, was a dodgy theorist and made a hash of trying to explain why.

Enter Einstein. In 1905, Einstein wrote a paper on the photoelectric effect. In this paper, he referred to work by the German physicist Max Planck, who five years earlier had drawn first blood in the tussle between classical Newtonian physics and the soon-to-be-formulated quantum mechanics. Planck was trying to explain the behavior of certain types of objects called black bodies, which are idealized objects in thermal equilibrium that absorb all the infalling radiation and radiate it back out. If the electromagnetic energy being emitted is infinitely divisible into smaller and smaller amounts, as it is in classical physics, thus making for a seamless continuum, then the predictions made by theory were at odds with experimental data. Something was not quite right with classical notions of energy.

To solve the puzzle, Planck argued that the spectrum of the black body electromagnetic radiation could be explained only if one thought of energy as coming in quanta, which are the smallest units of energy. Each unit is a quantum, and this quantum is a floor: for a given frequency of electromagnetic radiation, you cannot divide the energy into packets any smaller (the way you cannot divide a dollar into anything smaller than a cent). Using this assumption, Planck beautifully explained the observations. The idea of the quantum was born.

While Einstein did not fully embrace Planck's ideas in his 1905 paper to explain the photoelectric effect, he would eventually do so. Einstein argued that since light is electromagnetic radiation, it too comes in quanta: the higher the frequency of the light, the higher the energy of each quantum. This relation is linear—doubling the frequency doubles the energy of the quantum. Einstein's claim about light coming in quanta was crucial to understanding the photoelectric effect, in which light falling on a metal can sometimes dislodge an electron from an atom of the metal. For any given metal, said Einstein, an electron can be freed from the metal's surface only if the incident quantum of light has a certain minimum amount of energy: anything less, and the electrons stay put. This explains why electrons never leave the metal surface if the incident light is below a threshold frequency: the quantum of energy is too low. And it does not matter if two quanta put together have the necessary amount of energy. The interaction between light and an atom of metal happens one quantum at a time. So, just pumping more and more quanta below the threshold frequency has no effect.

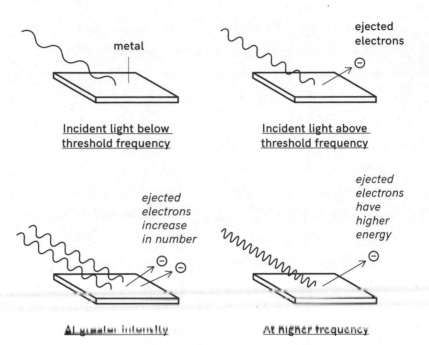

Incident light below
threshold frequency

Incident light above
threshold frequency

ejected
electrons

metal

ejected
electrons
increase
in number

ejected
electrons
have
higher
energy

At greater intensity

At higher frequency

With this theory, Einstein also predicted that the ejected electrons will get more energetic (or have greater velocities) as the frequency of the incident light increases. There is more energy in each quantum of light, and this imparts a stronger kick to the electron, causing it to fly out of the metal at a greater speed—a prediction that would soon get experimentally verified.

Einstein's profound claim here was that light is made of small, indivisible particles, where the energy of each particle or quantum depends on the frequency or color of the light. The odd thing, of course, is that terms like *frequency* and *wavelength* refer to the wave nature of light, and yet these were getting tied to the idea of light as particles. A disturbing duality was beginning to raise its head. Things were getting confusing.

Both Lenard and Einstein got Nobel Prizes for their work, Lenard in 1905 for his "work on cathode rays," and Einstein in 1921, for explaining the photoelectric effect using Planck's quantum hypothesis. Lenard, however, became deeply resentful of the accolades given to Einstein for theorizing about what Lenard regarded as his result. Lenard was an anti-Semite. In 1924, he became a member of Hitler's National Socialist party. In front of his office at the Physics Institute in Heidelberg, there appeared a sign that read: "Entrance is forbidden to Jews and members of the so-called German Physical Society." Lenard viciously attacked Einstein and his theories of relativity, with undisguised racism and anti-Semitism. "Einstein was the embodiment of all that Lenard detested. Where Lenard was a militaristic nationalist, Einstein was a pacifistic internationalist . . . Lenard decided that relativity was a 'Jewish fraud' and that anything important in the theory had been discovered already by 'Aryans,'" Philip Ball wrote in *Scientific American*.

In the midst of terrible social unrest and unhinged ideologies across Europe, the quantum revolution was set in motion.

As things stood in 1905, electrons were constituents of atoms (but it was still unclear whether that was the full story about the makeup of atoms). Plus, there were electromagnetic fields, which were described by Maxwell's equations. These came in quanta. It was clear that light too is an electromagnetic wave, which came in quanta and these quanta could be thought of as particles. Microscopic reality did not make a whole lot of sense.

J. J. Thomson, meanwhile, had a question he wanted answered.

What would happen when a few quanta of light went through a single slit (rather than two slits)? In 1909, a young scientist named Geoffrey Ingram Taylor started working with Thomson at his laboratory in Cambridge. Taylor decided to design an experiment to try and answer Thomson's question. The answer resonates within quantum mechanics even today, and is particularly relevant for the story of the double-slit experiment.

Picture a source of light that shines on an opaque sheet with a single slit. On the other side of the opaque sheet is a screen. Again, our naive expectation is that we'll see a single strip of light on the screen. Instead, what appear are fringes (albeit a different pattern than seen with the double slit. In the case of a single slit, the fringes can be explained by thinking of each point in the opening or aperture of the slit as a source of a new wave. These waves then interfere with each other, leading to what's called a diffraction pattern). It's another proof that light behaves like a wave. When there's lots of light, the results are easy to explain: light is an electromagnetic wave, and so we should see fringes.

But given that light also comes in quanta, or particles, Thomson wanted to understand the single-slit phenomenon when the intensity of light falling on the single slit is turned way down, so that only a few quanta of light go through the slit at any one time. Now, if the screen on the far side is a photographic plate that records each quantum of light, then over time, would one see interference fringes? Thomson argued that there should be blurry fringes, because in order to get sharp fringes, numerous quanta should arrive simultaneously at the screen and interfere. Reducing the quanta reaching the screen at the same time to a trickle should reduce the amount of interference and hence the sharpness of the fringes, Thomson hypothesized.

Taylor was in his twenties and starting out on his career as an experimental physicist. He chose this experiment as the subject of his first scientific paper, but oddly, he recalled years later, "I chose that project for reasons which, I fear, had nothing to do with its scientific merits." Consequently, he performed the experiment in the children's playroom of his parents' home. To create a single slit, he stuck metal foil onto a piece of glass and, using a razor blade, etched a slit in the metal foil. For a source of light he used a gas flame. Between the flame and the slit, he placed many layers of darkened glass. Taylor calculated that the light falling on the single slit was so faint that it was equivalent to a candle burning a mile away. On the other side of the slit, Taylor placed a needle, whose shadow he captured on a photographic plate. The light—ostensibly just a few quanta at a time—passed through the slit and landed on the photographic plate. What would the plate record after weeks of exposure to the faint light?

Taylor's mind, meanwhile, was elsewhere. He was becoming an accomplished sailor. He set up his experiment so that he could get enough of an exposure on the photographic plate after six weeks. "I had, I think rather skillfully, arranged that this stage would be reached about the time when I hoped to start a month's cruise in a little sailing yacht I had recently purchased," he said. During the longest stretch of the experiment, in which the photographic plate was exposed for three months, Taylor reportedly went away sailing.

After that three-month-long exposure, Taylor saw interference fringes—as sharp as if the photographic plate had been exposed to more intense light for a very short time. Thomson was proved wrong. Taylor never followed up on this negative result. If he had, he might have played an important role in the development of quantum mechanics—for his results were hinting at the odd behavior of photons. Instead of pursuing this any further, Taylor changed directions and went on to make seminal contributions to other fields of physics, particularly fluid mechanics.

Thomson, however, wasn't done being a mentor. In the autumn of 1911, a young Danish scientist named Niels Bohr came to work with Thomson. Soon thereafter, Bohr moved to Manchester to study with New Zealand–born British physicist Ernest Rutherford, who was probing the structure of the atom. Rutherford's work had established that the atom, besides having electrons, also has a positively charged nucleus. Calculations showed that much of the mass of the atom is in the nucleus. What emerged was a new picture of an atom: negatively charged electrons orbiting a positively charged nucleus, the way planets orbit the sun.

Almost immediately, physicists realized that this model had serious shortcomings. Newton's laws mandated that orbiting electrons had to be accelerating, if they were to remain in their orbits without falling into the nucleus. And Maxwell's equations showed that accelerating electrons should radiate electromagnetic energy, thus lose energy and eventually spiral into the nucleus, making all atoms unstable. Of course, that's not what happens in nature. The model was wrong.

An interim solution came courtesy of the young Bohr. In 1913, Bohr proposed that the energy levels of electrons orbiting a nucleus did not change in a continuous manner, and also that there was a limit to the lowest energy level of an electron in an atom. Bohr was arguing that the orbits of the electrons and hence their energy levels were quantized. For any given nucleus, there's an orbit with the lowest possible energy. This orbit would be stable, said Bohr. If an electron were in this lowest-energy orbit, it could not fall into the nucleus, because to do so, it'd have to occupy even smaller orbits with lower and lower energies. But Bohr's model prohibited orbits with energies smaller than the smallest quantum of orbital energy. There was nowhere lower for the electron to fall. And apart from this stable, lowest-energy orbit, an atom has other orbits, which are also quantized: an electron cannot go from one orbit to another in a continuous fashion. It has to jump.

To get a sense for how weird it must have been for physicists in the early twentieth century to understand Bohr's ad hoc claims, imagine you are driving your car and want to go from 10 to 60 miles per hour. In an analogy to the way electrons behave in orbits, the car jumps from 10 mph to 60 mph in chunks of 10 mph, without going through any of the intermediate speeds. Moreover, no matter how hard you brake, you cannot slow the car down to below 10 mph, for that's the smallest quantum of speed for your car.

Bohr also argued that if an electron moves from a high-energy to a low-energy orbit, it does so by emitting radiation that carries away the difference in energy; and to jump to an orbit with higher energy, an electron has to absorb radiation with the requisite energy.

To prevent electrons from losing energy while orbiting the nucleus, which they would have to according to Maxwell's theory,

Bohr argued that the electrons existed in special "stationary" states, in which they did not radiate energy. The upshot of this somewhat arbitrary postulate was that another property of electrons, their angular momentum, was also quantized: it could have certain values and not others.

It was all terribly confounding. Nonetheless, there were connections emerging between the work of Planck, Einstein, and Bohr. Planck had shown that the energy of electromagnetic radiation was quantized, where the smallest quantum of energy (E) was equal to a number called Planck's constant (h) multiplied by the frequency of the radiation (v), producing his famous equation $E=hv$. Einstein showed that light came in quanta, and the energy of each quantum or photon was also given by the same equation, $E=hv$ (where v refers to the frequency of the light).

While Bohr had shown that the energy levels in atoms were quantized, it'd take him a decade or so more to accept that when electrons jumped energy levels, the radiation going in or coming out of the atom was in the form of quanta of light (Bohr initially insisted that the radiation was classical, wavelike).

But when Bohr did accept Einstein's idea of light quanta, he saw that the absorbed or emitted energy of the photons was given by, again, $E=hv$. (Bohr wasn't the only big name resisting Einstein's ideas. The notion of light being quantized was hard to stomach for physicists, given the success of Maxwell's equations of electromagnetism in describing the wave nature of light. For instance, Planck, when he was enthusiastically recommending Einstein for a seat in the Prussian Academy of Sciences in 1913, slipped in this caveat about Einstein: "That he might sometimes have overshot the target in his speculations, as for example in his light quantum hypothesis, should not be counted against him too much.")

Still, the evidence for nature's predilection for sometimes acting like waves and sometimes like particles continued to grow. In 1924, Louis de Broglie, in his PhD thesis, extended this relationship to particles of matter too, and provided a more intuitive way to envision why the orbits of electrons are quantized. Matter, said de Broglie, also exhibited the same wave-particle duality that Einstein had shown for light. So an electron could be thought of as both a wave and a particle. And atoms too. Nature, it seems, did not discriminate: everything had wavelike behavior and particle-like behavior.

The idea helped make some sense of Bohr's model of the atom. Now, instead of thinking of an electron as a particle orbiting the nucleus, de Broglie's ideas let physicists think of the electron as a wave that circles the nucleus, the argument being that the only allowed orbits are those that let the electron complete one full wavelength, or two, three, four, and so on. Fractional wavelengths are not allowed.

It was clear by then that physics was undergoing a profound transformation. Physicists were beginning to explain previously inexplicable phenomena, using these ideas of quantized electromagnetic radiation, quantized electron orbits, and the like, at least for the simplest atom, that of hydrogen, which has one electron orbiting the nucleus. More complex atoms were not so easily tamed, even with these new concepts. Still, what was being explored was the very structure of reality—how atoms behave and how the electrons inside atoms interact with the outside world via radiation, or light. But the successes notwithstanding, the puzzles were also mounting.

While nature's discontinuity and discreteness at the smallest scales was becoming ever more obvious, there was the puzzling issue of its concomitant wave nature, emblematic of classical continuity. And, probably most disturbingly, there was the question of indeterminism. It was clear that there are natural phenomena that do not follow the clockwork determinacy of Newton's classical world. Take, for example, radioactivity: nothing about the current state of a radioactive atom lets you predict exactly when it'll emit a ray of radioactivity. The process is unpredictable, stochastic. This went against the tenets of the science of the time, according to which full knowledge of a system should let you predict with precision some future event involving that system. The microscopic world seemed to be operating with a different set of rules.

But it wasn't obvious what these rules were. What physics lacked was an overarching framework that brought these disparate elements together. All that changed during the mid to late 1920s, when in a few feverish years, brilliant minds forged not one but two frameworks for theorizing about the world at small scales. This effort would culminate in one of the most celebrated scientific conferences in history—the Fifth Solvay International Conference on Electrons and Photons held in October 1927 in Brussels, Belgium. The moment, captured in a now-iconic photograph taken by Belgian photographer Benjamin Couprie, shows all twenty-nine attendees, some standing in the back row still in their twenties and yet to become famous, some already so and seated in the front row, including Einstein, Planck, and Marie Curie, and almost everyone else in between who mattered to the emerging field of quantum physics. If they weren't already Nobel Prize winners, many would go on to win—turning seventeen of the twenty-nine into Nobel laureates.

"The lakes" of Copenhagen are five reservoirs that stretch crescent-shaped not far from the city center. Walk along the northern end of these lakes, go past a stretch of shore lined with horse chestnut trees, down a couple of blocks along an alley named Irmingersgade, and you come up, quite suddenly, on an unassuming building: the Niels Bohr Institute. When it was founded in 1921 by Bohr, it was called the Institute of Theoretical Physics. Bohr had moved from Manchester to the University of Copenhagen, where he became a professor in 1916 at just thirty-one years of age. He then lobbied hard and got the funds to build an institute for theoretical physics. And for a few decades, the institute became a cauldron where great minds stewed over the evolving field of quantum physics, under Bohr's deeply engaged gaze.

One of these great minds was a young German physicist named Werner Heisenberg. Bohr first met Heisenberg at Göttingen, Germany, in June 1922. Bohr was there to talk about the current understanding of the model of the atom and the various outstanding problems yet to be solved. During the talk, Heisenberg, still a twenty-year-old student in his fourth semester, questioned Bohr with such clarity that a suitably impressed Bohr took Heisenberg for a walk afterward to discuss atomic theory. He also invited Heisenberg to Copenhagen, and it was there in 1924 that Heisenberg realized, after discussions with Bohr and others, that "perhaps it would be possible one day, simply by clever guessing, to achieve the passage to a complete mathematical scheme of quantum mechanics." The word *mechanics* refers to physics that can explain how something changes with time under the influence of forces.

Heisenberg's insight was prophetic. In the spring of 1925, suffering from severe hay fever, he decamped to Helgoland in the North Sea, a rocky island devoid of pollen. There, between long walks and contemplating Goethe's *West-östlicher Divan*, he developed the early mathematics that would become the basis for modern quantum theory. Heisenberg recalled later, "It was almost three o'clock in the morning before the final result of my computations lay before me . . . I could no longer doubt the mathematical consistency and coherence of the kind of quantum mechanics to which my calculations pointed. At first, I was deeply alarmed. I had the feeling that, through the surface of atomic phenomena, I was looking at a strangely beautiful interior, and felt almost giddy at the thought that I now had to probe this wealth of mathematical structures nature had so generously spread out before me. I was far too excited to sleep, and so, as a new day dawned, I made for the southern tip of the island, where I had been longing to climb a rock jutting out into the sea. I now did so without too much trouble, and waited for the sun to rise."

Heisenberg wrote up his work, showed it first to Wolfgang Pauli (another of the brilliant young minds) and then to Max Born (an equally brilliant but a more fatherly figure in his forties, with whom Heisenberg was doing his postdoctoral work). Born immediately realized the import of Heisenberg's paper. "I thought the whole day and could hardly sleep at night . . . In the morning I suddenly saw the light," he would say.

What Born realized was that the symbols Heisenberg was manipulating in his equations were mathematical objects called matrices, and there was an entire field of mathematics devoted to them, called matrix algebra. For example, Heisenberg had found that there was something strange about his symbols: when entity A was multiplied by entity B, it was not the same as B multiplied by A; the order of multiplication mattered. Real numbers don't behave this way. But matrices do. A matrix is an array of elements. The array can be a single row, a single column, or a combination of rows and columns. Heisenberg had brilliantly intuited a way of representing the quantum world and asking questions about it using such symbols, while being unaware of matrix algebra.

In a few frenetic months, Born, along with Heisenberg and Pascual Jordan, developed what's now known as the matrix mechanics formulation of quantum physics. In England, Paul Dirac saw the light too when he encountered Heisenberg's work, and he too, in a series of papers, independently added tremendous insight and mathematics to the formulation and developed the "Dirac notation" that's still in use today.

Most important, it was clear that the formalism worked. For example, the position of, say, an electron, is represented by a matrix. The position in this case is called an observable. The matrix then dictates all the possible positions in which the electron can be found, or observed. The formalism implicitly allows for the electron to be only in certain positions and not in others. And there is no sense of a continuous change from one position to another. Discreteness, or jumps from one state to another, is baked into matrix mechanics.

In due course, physicists were able to use the formalism to calculate, for example, the energy levels of electrons in atoms, explain the radiation emitted by glowing bits of sodium or other metals, understand how such spectral emissions could be split into slightly different frequencies under the influence of a magnetic field, and better understand the hydrogen atom itself.

But it wasn't obvious why the formalism worked. What did these matrices map to, physically speaking? The elements of these matrices could be complex numbers (a complex number has a real part and an imaginary part; the imaginary part is a real number multiplied by the square root of -1 and is imaginary because $\sqrt{-1}$ doesn't exist yet turns out to be incredibly useful in certain kinds of mathematics). How could the physical world be represented by things that could only be imagined? Were we at the very limit of human understanding? Was a clear understanding possible?

Matrix mechanics does not allow physicists to think of electrons as having clear, fixed orbits, even if they are quantized. One can describe an electron's quantum state using a set of numbers, carry out a whole lot of matrix manipulations to predict things like spectral emissions, but what you lose is the ability to visualize the electron's orbit in the way that one can visualize, say, Earth's orbit around the sun.

Plus, the formalism deals in probabilities. If a particle is in state A and you measure to see if it's state A, then, of course, the math says you'll find the particle in state A with 100 percent certainty. The same goes for, say, state B. But matrix mechanics says that a particle can be in some intermediate state, where the state is x parts A and y parts B. Now, if you try and predict whether you'll find the particle in state A or state B, 100 percent certainty about reality is no longer possible.

Matrix mechanics lets you calculate only the probabilities of outcomes of measurements. So, for an electron whose state is x parts A and y parts B, say you want to see if the particle is in state A. The math says that the probability of finding the particle in state A is x^2. Similarly, if you check to see if the particle is in state B, the probability you'll find it in state B is y^2. (The terminology gets tweaked a little bit when you allow x and y to be complex numbers, but for now, it's easy to see what rules x and y have to follow: the probabilities have to add up to one, so $x^2 + y^2$ should equal 1.)

The fact that we are now dealing in probabilities is not, presumably, because we do not know enough about the particle. Matrix mechanics says you have all the information you can possibly have. Yet, if you take a million identically prepared particles in the same state (the same combination of states A and B) and perform a million identical measurements, then, on average, x^2 number of times you will find the particle in state A, y^2 of the time you'll find it in state B. But you can never predict the answer you'll get for any single particle. You can only talk statistically. Nature, it seems, is not deterministic in the quantum realm.

Recall that something similar happens with the double slit. We cannot predict where exactly a single photon will land on the screen—we can only assign probabilities for where it might go.

Soon after these phenomenal developments, an Austrian physicist named Erwin Schrödinger, whose status as a founding member of quantum physics was yet to be established, expressed his dismay at, even distaste for, Heisenberg's matrix mechanics. He said he was "discouraged, if not repelled" by what he saw as "very difficult methods of transcendental algebra, defying any visualization."

The battle lines were being drawn. Wave versus particle, continuous versus discrete, old versus new. Schrödinger's distaste led him to develop a formidable old-school alternative to the upstart, matrix mechanics—one that seemed to restore faith in the classical way of thinking about nature.

When Louis de Broglie wrote his 1924 thesis on the wave-particle duality of matter, Schrödinger was already a professor of theoretical physics at the University of Zurich, and compared to the young geniuses elsewhere in Europe, he was practically an old man, approaching forty. But for years Schrödinger had been delving into the same questions that had been tormenting everyone. Schrödinger learned of de Broglie's work when he read a reference to it in a paper by Einstein. Thinking of matter as waves made sense to Schrödinger's classically intuitive mind, and he acknowledged as much in a letter to Einstein, dated November 3, 1925: "A few days ago I read with the greatest interest the ingenious thesis of Louis de Broglie, which I finally got hold of." Schrödinger wanted to describe the motion of electrons around the nucleus by thinking of them as waves. Instead of Heisenberg's matrix mechanics, Schrödinger wanted wave mechanics for electrons.

If Heisenberg's solo sojourn at Helgoland has become quantum physics lore, so has Schrödinger's own burst of creativity in isolation—well, almost in isolation. A *New York Times* book review captures this period in Schrödinger's life: "A few days before Christmas, 1925, Schrödinger . . . took off for a two-and-a-half-week vacation at a villa in the Swiss Alpine town of Arosa. Leaving his wife in Zurich, he took along de Broglie's thesis, an old Viennese girlfriend (whose identity remains a mystery) and two pearls. Placing a pearl in each ear to screen out any distracting noise, and the woman in bed for inspiration, Schrödinger set to work on wave mechanics. When he and the mystery lady emerged from the rigors of their holiday on Jan 9, 1926, the great discovery was firmly in hand."

Within weeks, Schrödinger published his first paper in the *Annalen der Physik.* Three more papers followed in quick succession, and Schrödinger turned the world of Heisenberg and Born upside down. Suddenly, physicists had an intuitive way of understanding what was ostensibly happening to an electron in a hydrogen atom. Schrödinger had come up with his now eponymous wave equation, which treated the electron as a wave, and showed how this wave would change over time. It was wave mechanics. It was almost classical physics, except there were curious and consequential differences.

In classical physics, solving a wave equation for, say, a sound wave can give you the pressure of the sound wave at a certain point in space and time. Solving Schrödinger's wave equation gives you what's called a wavefunction. This wavefunction, denoted by the Greek letter ψ (psi, pronounced "sigh"), is something quite strange. It represents the quantum state of the particle, but the quantum state is not a single number or quantity that reveals, for example, that the electron is at this position at this time and at that position at another time. Rather, ψ is itself an undulating wave that has, at any given moment in time, different values at different positions.

Even more weirdly, these values are not real numbers; rather, they can be complex numbers with imaginary parts. So the wavefunction at any instant in time is not localized in a region of space; rather, it is spread out, it's everywhere, and it has imaginary components. The Schrödinger equation, then, allows you to calculate how the state of the quantum system, ψ, changes with time.

Schrödinger thought the wavefunction provided a way to visualize what was actually happening to electrons or other inhabitants of the quantum world. But this view was challenged within months of Schrödinger's papers being published, when Max Born realized that Schrödinger was wrong about the meaning of the wavefunction.

In a couple of seminal papers published in the summer of 1926, Born showed that when electrons collide and scatter, the resulting wavefunction that represents the state of the electrons only encodes the probability of finding the electrons in one state or another. It took Born a couple of tries to get it right, but he showed that if ψ is the wavefunction of an electron, and if it can be written, for example, in terms of two different possible states of the electron, ψA and ψB, such that $\psi = x.\psi A + y.\psi B$, then all you can do is calculate the probability that you'll find the electron in state A or state B when you do a measurement. (The probability of finding the electron in state A is given by the square of the amplitude of x, also called the square of the modulus of x, denoted as $|x|^2$, and the probability of finding it in state B is given by $|y|^2$. If x is, say, a real number, the modulus $|x|$ is simply its absolute value: if it's positive to start with, it remains positive; if it's negative, then we multiply it by -1; squaring it gives us a positive number. Of course, x and y can be complex numbers, and calculating the modulus of a complex number is a bit more complicated, but in essence, when you take the modulus-squared of a complex number, you again get a number that is positive and real, without any imaginary parts.)

Born had, it seemed at first blush, cast doubt on causality, the underpinning of deterministic classical physics, which says any given effect has a cause. Given an initial state of an electron, standard quantum mechanics cannot definitively say what the electron's next state will be. One can only calculate the probability of an electron transitioning to some new state, using what came to be called the Born rule. An element of randomness, or stochasticity, became an integral part of the laws of nature. As Born put it, "The motion of particles follows probability laws but the probability itself propagates according to the law of causality."

And there it was—one interpretation of the wavefunction. It's a probability wave. Schrödinger's equation lets you calculate how this wave changes with time deterministically, but as it evolves and takes on different shapes, what's changing are the probabilities of finding the quantum system in various states.

If this sounds like the probabilities of matrix mechanics, you are not mistaken. Schrödinger himself, in another stroke of insight, showed that wave mechanics and matrix mechanics are mathematically equivalent (in hindsight, it was a mathematician called John von Neumann who would really prove the equivalence a few years later). Rather than see this as a validation of matrix mechanics, Schrödinger claimed victory for wave mechanics, considering his approach to be correct and arguing that anything that was calculated using matrix mechanics could be calculated using wave mechanics. The advantage of wave mechanics, in Schrödinger's opinion, was the idea that nature even at the smallest scales was continuous, not discrete. There were no quantum jumps.

Heisenberg, meanwhile, wasn't enamored of Schrödinger's ideas. He wrote to Pauli, complaining that he found them "abominable," calling it *"Mist"* (which is German for rubbish, manure, dung, or droppings). Pauli himself alluded to *Züricher Lokalaberglauben* (local Zurich superstitions, an allusion to the city where Schrödinger worked). Schrödinger, unsurprisingly, wasn't pleased by Pauli's assertions. Pauli, in turn, tried to appease Schrödinger by saying, "Don't take it as a personal unfriendliness to you but look on the expression as my objective conviction that quantum phenomena naturally display aspects that cannot be expressed by the concepts of continuum physics. But don't think that this conviction makes life easy for me. I have already tormented myself because of it and will have to do so even more."

The torment these titans felt over the nature of reality continued when Schrödinger visited Copenhagen and met Bohr for the very first time.

Decades after Schrödinger's visit to Copenhagen in September 1926, Heisenberg would recount the intensity of their meetings: "The discussion between Bohr and Schrödinger began at the railway station in Copenhagen and was carried on every day from early morning till late at night. Schrödinger lived at Bohr's house so that even external circumstances allowed scarcely any interruptions of the talks. And although Bohr as a rule was especially kind and considerate in relations with people, he appeared to me now like a relentless fanatic, who was not prepared to concede a single point to his interlocutor or to allow him the slightest lack of precision. It will scarcely be possible to reproduce how passionately the discussion was carried on from both sides."

So passionately that even after Schrödinger fell sick and was bedridden with a fever and cold, the host did not relent. Bohr turned up at his bedside to debate quantum physics, even as Bohr's wife, Margrethe, took care of Schrödinger.

The debate between Bohr and Schrödinger was a foretaste of future debates that Bohr would have with Einstein about how to think about the smallest constituents of reality (at the time, electrons and photons). It was a clash of two ways of thinking. As Walter Moore writes in his book *Schrödinger: Life and Thought*, "Schrödinger was a 'visualizer' and Bohr was a 'nonvisualizer,' one thought in terms of images and the other in terms of abstractions."

Schrödinger left Copenhagen, but Heisenberg was still there to serve as Bohr's debating partner. Heisenberg was now living in an attic apartment at the institute, and it was there that Bohr would turn up late at night to continue their arguments. And though the two were mostly on the same side of the debate, they still had differences: Bohr wanted to make wave-particle dualism—the idea that nature has two faces and only shows one or the other at any one time—a key component of any interpretation of reality; Heisenberg put his "trust in the newly developed mathematical formalism," to see what meanings it suggested, rather than presupposing any particular view of reality.

They fretted about making sense of experiments, including the double slit. As Heisenberg would say, "Like a chemist who tries to concentrate his poison more and more from some kind of solution, we tried to concentrate the poison of the paradox, and the final concentration was such experiments like the electron with the two holes . . . They were just a kind of quintessence of what was the trouble."

By the end of February 1927, their discussions at an impasse, Bohr went off to ski in Norway. Heisenberg too took time for himself. He wrote of one extraordinary night when something clarified: "I went for a walk in the Fælledpark, which lies behind the institute, to breathe the fresh air and calm down before going to bed. On this walk under the stars, the obvious idea occurred to me that one should postulate that nature allowed only [those] experimental situations to occur which could be described within the framework of the formalism of quantum mechanics. This would apparently imply, as one could see from the mathematical formalism, that one could not simultaneously know the position and velocity of a particle."

Heisenberg had discovered the uncertainty principle. The formalism of quantum mechanics has pairs of observable quantities, such as the position and momentum of a particle, where trying to determine one with increasing precision means that you increase the imprecision of the values you obtain for the other. So, if you know exactly where a particle is, you have very little idea of its momentum, and vice versa. This relation extends to other pairs of quantities, such as energy and time.

(When I visited the Niels Bohr Institute, I went up to the attic to see Heisenberg's living quarters. His apartment was being used by builders to store air-conditioning equipment. A cartoon captioned "At home with the Heisenbergs" was stuck on the bathroom door outside the apartment, with Mrs. Heisenberg saying, "I can't find my car keys," and Mr. Heisenberg replying, "You probably know too much about their momentum.")

Bohr, meanwhile, became ever more convinced that what he called the principle of complementarity was at the heart of quantum mechanics: that wave nature and particle nature are complementary aspects of reality, and that it's our choice of experiment that reveals one or the other, but never both at the same time. He thought that the uncertainty principle was one outcome of the broader principle of complementarity.

Elsewhere, Einstein was growing deeply concerned about such interpretations of the quantum formalism, and building himself up toward a profound intellectual debate with Bohr, a debate that would shape the future of quantum mechanics. Einstein had a predilection for conjuring up thought experiments to make a point—and one of these involved the double-slit experiment. He brought it up at the Fifth Solvay Conference.

History has often portrayed Einstein and Bohr as giants in battle, slashing at each other with their respective intellectual might. But often what gets lost in the retelling is the enormous respect and affection that the two had for each other. Einstein and Bohr met for the first time in Berlin in April 1920. Impressed by Bohr, Einstein wrote to him in May, from America, beginning his letter with these words: "Dear Mr. Bohr: The magnificent gift from the neutral world, where milk and honey still flow, gives me a welcome occasion to write to you. Not often in life has a person, by his mere presence, given me such joy as you. I understand why [Paul] Ehrenfest is so fond of you." Bohr wrote back in June, saying, "To meet you, and talk with you, was one of the greatest experiences I have ever had."

This mutual admiration underpinned their relationship, despite their strong disagreements over quantum mechanics.

Their friendly salvos were fired in earnest at the Fifth Solvay Conference in Brussels. This was a grand battle of ideas, the likes of which occur infrequently enough in science to be etched in cultural memory as moments that changed our understanding of our place in the universe. Sometimes the individuals debating have been separated by the intervening centuries, as was the case with Copernicus, who in the sixteenth century argued against the Greek astronomer and mathematician Ptolemy's ancient theory that Earth is at the center of the solar system. Copernicus put the sun at the center. Sometimes, it's one person's fight against an emerging consensus, as was the case in the 1950s with English astronomer Fred Hoyle's increasingly isolated stand for a steady-state universe, when theory and evidence were both pointing to an expanding cosmos that began in a big bang. And at times, the antagonists debated the nature of scientific progress itself, as happened between philosophers Karl Popper and Thomas Kuhn. Popper, impressed by Einstein's work on relativity, argued that science progresses in increments; scientists come up with hypotheses to explain phenomena, hypotheses that they then try their best to falsify. Kuhn would be influenced by the goings-on at the Fifth Solvay Conference and argued that science mostly moves along in the manner suggested by Popper, with scientists working within an accepted paradigm, until anomalies—things that cannot be explained within the current way of thinking—pile up, bringing science to the brink of crisis, causing an upheaval and a dramatic paradigm shift.

The debates at the Fifth Solvay Conference set the stage for just such a shift. Bohr, Heisenberg, and Pauli were making a case for what eventually came to be called the Copenhagen interpretation of quantum mechanics. According to them, the only aspects of reality that you could know about were those that were allowed by the formalism. For example, you could ask about the probability of finding an electron somewhere, but you couldn't ask what path it took to get there, because there is nothing in the math that captures an electron's path. It'd take another five years for the math to become sophisticated, thanks to John von Neumann, but the new view of reality was taking hold. Taken at its most extreme, the Copenhagen interpretation is anti-realist: it denies any notion of reality that exists independent of observation. More important, the proponents were claiming that the mathematical formalism is complete, and that there is nothing more to say about reality.

This was, of course, a massive shift in our way of thinking. Until then, our theories said something concrete about a natural world that exists regardless of observation. Einstein, a realist, argued that the mathematical formalism of quantum mechanics was incomplete and did not paint a full picture of reality.

The Solvay Conference was being held at the Institute of Physiology in the heart of Brussels. "However, with all the participants staying at the Hotel Metropole, it was in its elegant art deco dining room that the keenest arguments took place . . . The acknowledged master of the thought experiment, Einstein would arrive at breakfast armed with a new proposal that challenged the uncertainty principle and with it the much-lauded consistency of the Copenhagen interpretation. The analysis would begin over coffee and croissants. It continued as Einstein and Bohr headed to the Institute of Physiology, usually with Heisenberg, Pauli and Ehrenfest trailing alongside. As they walked and talked, assumptions were probed and clarified before the start of the morning session . . . During dinner back at the Metropole, Bohr would explain to Einstein why his latest thought experiment had failed to break the limits imposed by the uncertainty principle. Each time Einstein could find no fault with the Copenhagen response, but they knew, said Heisenberg, 'in his heart he was not convinced.'"

At the center of one of their mind games was the double-slit experiment. Einstein imagined an electron that first passes through a single slit, and then encounters a double slit, and eventually ends up somewhere at the center of the far screen. In Einstein's original

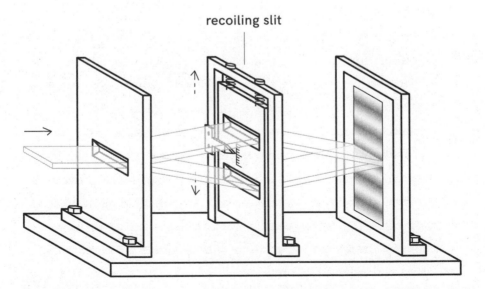

recoiling slit

thought experiment, the single slit could move up and down, while the double slit was fixed, but physicists since then have reimagined the setup with the single slit held in place, and the double slit as the one that can move up or down as it's buffeted by the particles going through the slits. While conceptually identical to Einstein's imagined apparatus, the newer version is easier to grasp.

Consider an electron that goes through the single slit, then through the double slit, and then lands at the center of the far screen. Using Einstein's analysis, if the electron went through the lower slit, then it'd have had to change directions and move upward to get to the center of the screen. This would impart a downward kick to the slit itself. And if the electron went through the upper slit, it'd impart an upward kick to the slit. So, by measuring the momentum transfer, one should be able to tell which slit the electron went through, said Einstein. His point was that even though one observes the interference pattern, which demonstrates the electron's wave nature, measuring the slit's momentum tells us about the electron's path on its way to the far screen, thus revealing its particle nature. The two aspects of reality are not mutually exclusive, he claimed, and the fact that quantum mechanics did not have the formalism to capture that fact meant that it was somehow incomplete.

Bohr was stumped for a bit, but soon came back with a retort (in addition to coming up with the drawings that involved bolting the apparatus to a base and other practical things). He pointed out that if the slit can move when the electron passes through, and if we can measure the momentum transfer with precision, then we'll have imprecise knowledge about its location (thanks to Heisenberg's uncertainty principle). Now, if you do the calculations of where the electrons land on the far screen, taking into account the uncertainty about the slit's position, it turns out the interference pattern gets smudged. Trying to find out which slit the electron went through, by allowing the slits to move, destroys its wave nature. We can see the electrons either as particles or as waves, not both at the same time.

This was, of course, a thought experiment. There was no way to implement such an exquisitely engineered experiment in the 1920s, to get information about the particle's path without destroying the particle. It'd take almost a century of effort to carry out a variation of this thought experiment. It turns out that Bohr was right in this regard: it's impossible to dupe nature. (However, physicists and historians reading Bohr's writings would point out later that Bohr's arguments were somewhat inscrutable, so one should be circumspect about unqualified claims that "Bohr was right"—nonetheless, as experimental evidence goes, it went against Einstein on this count.) The experiment also showed that complementarity is a seemingly more powerful principle than maybe even Bohr imagined.

Such victories in hand, Bohr and company started giving concrete shape to the Copenhagen interpretation and its anti-realist view of nature. In the double-slit experiment, the Copenhagen interpretation makes no claim as to the path of the particle through the apparatus and, some would say, even denies that such a path exists.

Einstein and Bohr continued sparring over what quantum mechanics was telling us about reality. Was quantum physics the whole story? Was the mathematical formalism that described the statistical behavior of the subatomic world a complete description of reality? Or was there a hidden reality that the math wasn't capturing? Bohr metaphorically shrugged his massive shoulders and insisted there was no hidden reality.

Bohr, for his part, kept returning to the double-slit experiment to make philosophical points, sometimes infuriating his audience. Hendrik Casimir, a young physicist who had come to work with Bohr, wrote about a conversation with Bohr and Danish philosophers Harald Høffding and Jørgen Jørgensen. They were all at the Carlsberg mansion (the erstwhile residence of the founder of the Carlsberg brewery). Bohr was talking about the double-slit experiment done with electrons. Someone quipped, "But the electron must be somewhere on its road from source to observation screen." Bohr pointed out that the answer depends on what one means by the phrase *to be*. An exasperated Jørgensen retorted: "One can, damn it, not reduce the whole of philosophy to a screen with two holes."

But Bohr wasn't being flippant. What does it mean *to be* something in the quantum realm? Opinions differ dramatically. And the experiment with two holes, despite Jørgensen's protestations, remains at the center of these historic, differing scientific and philosophical arguments.

3

BETWEEN REALITY
AND PERCEPTION

Doing the Double Slit, One Photon at a Time

> The electron, as it leaves the atom, crystallises out of
> Schrodinger's mist like a genie emerging from his bottle.
>
> —Arthur Eddington

The 2014 Nobel Prize winner for chemistry, Stefan Hell, during his Nobel banquet speech, recalled the 1933 Nobel Prize winner Erwin Schrödinger as saying, "It is fair to state that we are not going to experiment with single particles any more than we will raise dinosaurs in the zoo."

Hell, speaking eighty-one years after Schrödinger's comment, quipped, "Now, ladies and gentlemen, what do we learn from this? First. Erwin Schrödinger would never have gone on to write *Jurassic Park* . . . Second. As a Nobel Laureate you *should* say 'this or that is *never* going to happen,' because you will increase your chances tremendously of being remembered decades later in a Nobel banquet speech."

The reference to dinosaurs and Schrödinger's skepticism of single-particle experiments came up when I met Alain Aspect, a French experimental physicist at the Institut d'Optique in Palaiseau, a suburb of Paris. Aspect, in fact, pioneered some of the first experiments done with single photons, including the first-ever double-slit experiment done by sending single photons through the apparatus. It was a pivotal moment in the more than half-a-century-long story of quantum physics, one that gave credence to all the theorizing that had come before, while laying the foundation for similar, more sophisticated experiments to come.

When we met more than twenty-five years after he had announced the results of his pioneering experiments, Aspect spoke as an elder statesman of quantum physics; the combination of his French-accented English and luxurious, graying mustache reminded me of the fictional detective Hercule Poirot in Agatha Christie mysteries (with apologies to Aspect; Poirot, of course, is Belgian).

In the early 1970s, Aspect finished his master's degree and went to Cameroon, Africa, to teach schoolchildren, as part of his mandatory French military service. While in Cameroon, his mind was on physics. He couldn't shake the feeling that there was something lacking in what he had learned. All the physics he had been taught—things like optics, electromagnetism, and thermodynamics—dealt with the classical, continuous, and deterministic world of Newton, Maxwell, and Einstein. He knew little about the physics of the microscopic quantum world of particles and atoms. When he heard talk of how an atom jumps from one energy level to another by emitting or absorbing a photon of light, he couldn't understand how. "I knew I was missing something," he said.

So, Aspect bought a newly published book, simply titled *Quantum Mechanics* (one that would become a highly regarded textbook, and one of the authors of which, Claude Cohen-Tannoudji, would become Aspect's PhD thesis advisor and would also win the Nobel Prize in 1997). Aspect read the book from cover to cover, or as he put it, "from page one to, I don't know, page 1,300." He was hooked.

When he returned to France in 1974, Aspect came upon a decade-old paper by John Bell, a Northern Irish physicist working at CERN (European Organization for Nuclear Research), the particle physics lab near Geneva, Switzerland. Bell wasn't yet famous for his 1964 paper, which contained a theorem that is now regarded as his signature contribution. When Aspect read the paper, which he did in a two-hour sitting, he said to himself, "This is unbelievable . . . it's fantastic." He realized that Bell's paper offered a way of resolving an argument over the nature of reality that had so consumed Einstein and Bohr (others had realized it too, but for young Aspect it was a revelation).

Bell's '64 theorem made it possible to experimentally address the question posed by Einstein: were there *local hidden variables* to describe properties of quantum systems that weren't there in the standard formalism of quantum mechanics, variables that in Einstein's opinion would turn quantum mechanics into a complete description of reality? The word *local* refers to elements of reality that cannot influence each other any faster than the speed of light: *local variables* are bits of mathematics in our theories that represent this reality, and *hidden local variables* refer to variables that are, well, absent from the formalism. Bell was particularly interested in locality. While some physicists had already done experiments based on Bell's ideas, the results weren't conclusive. In Aspect's eyes, these experiments had failed to come close to the ideal experiment demanded by Bell's theorem. He felt he could do a better job.

But Aspect had some doubts. For one, he had yet to start on his PhD. Was this the right project for a doctoral thesis? Aspect went to CERN to meet Bell for advice, and Bell reassured the young Aspect that he was on the right track. But Bell also warned Aspect that the topic was considered by many as "crackpot physics." Almost no one doubted the completeness of quantum mechanics. So why bother testing it? Bell, worried for the Frenchman's career, asked Aspect if he had a permanent job. "I did. It was a small position, but it was permanent," Aspect recalled telling Bell. "They could not fire me, I was sure to get a salary each month."

Aspect returned to France and embarked on an experiment that is now regarded as the first to all but rule out a class of so-called hidden reality theories. To accomplish his feat, Aspect developed the technology to generate single particles of light, or photons, which could be sent into his apparatus one at a time. The use of single photons caught Richard Feynman's attention. In 1984, Aspect was invited to Caltech to speak about his tests of Bell's theorem. Feynman was in attendance. "Everybody expected to see Feynman destroy this young French guy pretending to settle a question that didn't exist," said Aspect.

During the Q&A session that followed Aspect's talk, Feynman amiably asked if Aspect could use single photons to perform an older, more classic experiment in quantum mechanics, one that

Feynman had highlighted in his own talks and lectures as one that best probed the mysteries at the heart of quantum mechanics: the double-slit experiment done with single particles. Aspect respectfully replied that a student of his, Philippe Grangier, was on the case back in Paris.

From the time Young did his sunbeam experiment in 1801 and through the development of quantum mechanics, no one had actually done the double-slit experiment with single photons. Until Aspect came along, no one even knew how to generate single photons *and* be sure that there was only one photon in the apparatus at any given time. "Ordinary sources do not emit well-separated single photons. In a discharge lamp, in a bulb, or even in a laser, you have always zillions of atoms simultaneously emitting photons," said Aspect. "As a result, what you get is an ensemble of photons, and this ensemble of photons has all the properties that can be described by classical electromagnetic waves."

For example, we now know that Geoffrey Ingram Taylor's detection of interference fringes, which was done with light so faint that it was like placing a candle a mile away, was not the result of single photons hitting the photographic plate. Taylor used something called a coincidence detector, which requires at least four photons to hit the detector at the same time to create a signal big enough to be recorded.

To appreciate how difficult it is to get photons one at a time from a source of light, consider this: you have a 100-watt lamp and you are monitoring the number of photons reaching a square opening, one centimeter to a side, placed one meter from the lamp. A rough-and-ready calculation (by Giancarlo Ghirardi in *Sneaking a Look at*

God's Cards) reveals that about 24 million billion photons will pass through that one-square-centimeter opening every second. That's 24 quadrillion photons. Getting single photons would require a technique substantially different from just turning down a lamp or dimming the light from a candle. Aspect developed such a technique. And when he carried out the double-slit experiment with single photons, classical physics had no say; only quantum mechanics could explain the results.

While single photons proved difficult to tame until Aspect came along, physicists weren't biding their time, waiting for technology to catch up. There were other particles to work with. Recall that Feynman had focused on the double-slit experiment done with single electrons. He emphasized, however, that it was simply a thought experiment. In 1961–1962, Feynman gave a series of lectures to freshmen and sophomores at Caltech, which were published a year later as a three-volume set. In it, he said about single electron interference: "This experiment has never been done in just this way. The trouble is that the apparatus would have to be made on an impossibly small scale to show the effects we are interested in." Feynman had no way of knowing that the impossible had been accomplished in 1961—in Germany, the results being published in German.

The 1961 experiment had its roots in work done by Gottfried Möllenstedt at the University of Tübingen. Möllenstedt invented a unique device to split a beam of electrons into two and then get them to interfere, analogous to Thomas Young using a thin card to split his sunbeam. The device is called an electron biprism. Möllenstedt came upon the idea accidentally. In the early 1950s, he was using an electron microscope, with a thin tungsten wire stretched across the objective lens of the microscope. Möllenstedt noticed that when the tungsten wire developed a charge, this caused two images to be formed, as if the microscope was seeing double. Möllenstedt realized that the charged wire was causing the microscope's electron beam to part, creating the two images. Could such a charged wire be used to see interference fringes by splitting an electron beam and letting it recombine?

Möllenstedt and his student Heinrich Düker took on the task. For a thin wire, they initially used gold-plated strands of spider silk (apparently, Möllenstedt "kept a collection of spiders around the laboratory for this purpose"). Eventually, the duo figured out how to gold-plate quartz wires only about 3 microns in diameter (human hair, for comparison, is about 100 microns thick). They charged the wire by applying a voltage to it, and placed the wire in the path of a beam of electrons. The electrons, deflected by the wire's charge, curved around the wire and eventually recombined. This was conceptually identical to a double-slit experiment: the electrons could take one of two paths, as they would if they encountered two slits.

However, despite using "powerful optics," the team did not see any fringes at first—they were just too small, just as Feynman had feared. But when they exposed a photographic plate to the electrons for 30 seconds and then looked at the photograph, using a powerful optical microscope, they saw "fine interference fringes." This was in 1954. In a paper published soon after, in *Naturwissenschaften*, they compared their fringes to optical fringes seen earlier by the French physicist Augustin-Jean Fresnel. The journal's editor, while praising

Möllenstedt and Düker, also pointed out that "Thomas Young had produced [such] . . . interferences ten years before Fresnel."

It's easy to understate the importance of this work. Electrons—which are particles of matter—are producing interference fringes, which are a phenomenon associated with waves. This was exactly what Louis de Broglie had postulated in 1924: that matter, and not just light, exhibited wave-particle duality. Using nothing but the value of the voltage applied to the quartz wire, the geometry of the apparatus, and the observed fringes, Möllenstedt and Düker proved correct de Broglie's formula for matter waves, which says that the wavelength of a particle, λ, equals Planck's constant, h, divided by the particle's momentum, p (so $\lambda = h/p$). The equation is audacious: the left-hand side deals with the properties of waves, and the right-hand side with properties of particles. You couldn't ask for a more succinct expression of wave-particle duality. Möllenstedt wrote to de Broglie, and de Broglie replied, "It was . . . a great pleasure to see that you have obtained a new and particularly brilliant proof of the formula."

Möllenstedt's young German student, Claus Jönsson, watched these experiments being done. By 1961, Jönsson performed a formal double-slit experiment with electrons, the same year that Feynman started talking of his thought experiments at Caltech. Jönsson wrote up his reports in German; it'd take years for the work to be translated into English, explaining why Feynman continued to regard it as an experiment in thought only.

But these experiments thus far were done with many electrons passing through the double slit (or passing by the charged wire) at the same time. The experiment with single electrons took somewhat longer. And different teams claim the credit for pulling it off: one in 1974, and the next in 1989. Conceptually, the experiments were similar to Möllenstedt and Düker's, except for the guarantee that only one electron was going through the apparatus at any one time (whether this actually happened is the source of dispute between the two teams).

In 1974, Italian physicists Pier Giorgio Merli, GianFranco Missiroli, and Giulio Pozzi in Bologna, Italy, recorded on a television monitor the arrival of the electrons after they had gone past the biprism. To observe the interference pattern with the naked eye required some fancy optics that magnified the fringes a few hundred times. The Italians also had to develop technology for "storing" the electrons that arrived at the monitor for a few minutes so that one could see the fringes once all the electrons had been collected otherwise the spot created by the first electron would have long vanished by the time the last electron arrived. The team made a 16mm movie of the fringes taking shape on the monitor; the movie even won an award at the Seventh International Scientific and Technical Movie Festival held in Brussels in 1976.

In 1989, Akira Tonomura and colleagues at Hitachi in Japan did a similar experiment, with an extremely well controlled source of electrons. Tonomura's team also created technology to record electrons on a screen the way one can record particles of light on a photographic plate: one by one, to build up an image over time. This way, they didn't have to store the first electron until the last one arrived. The Hitachi team's film of the electrons hitting the screen (the actual elapsed time was twenty minutes, but the film is sped up) is one of the most fascinating short films in the history of physics.

Electrons appear as dots on the screen, seemingly at random, but soon enough the fringes build up, a magical demonstration of single-particle interference. The movie belies the challenge the experimenters faced: the entire equipment, including the electron source and the biprism, had to be held unerringly stable the entire time; had anything moved even fractions of a micron, it would have destroyed the fringes.

A little over a decade later, *Physics World* published an article to celebrate the fact that the double-slit experiment done with single electrons was voted "the most beautiful experiment" in physics. It failed to mention the 1974 Italian effort, prompting a letter of protest from the Italian team. *Physics World* updated its article, including the Italian team's letter and Tonomura's reply arguing for the Japanese team's place in history: "We believe that we carried out the first experiment in which the build-up process of an interference pattern from single electron events could be seen in real time as in Feynman's famous double-slit *gedankenexperiment*. This was under the condition, we emphasize, that there was no chance of finding two or more electrons in the apparatus."

There is, however, no doubt about who was the first to test the double slit with single photons.

The double-slit experiment that Aspect and Grangier did in Paris starts with a piece of glass that reflects half the light that falls on it, and transmits the other half. This is actually a fairly common occurrence with glass. Think of sitting in a train that's racing through the countryside at night. If it's completely dark outside and you are looking at the windowpane, you will see the inside of the carriage

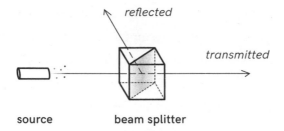

reflected

transmitted

source   beam splitter

reflected in the glass. But when the train passes by some lighted buildings outside, you see the buildings while simultaneously seeing your own reflection in the glass. The windowpane is both reflecting and transmitting light. In labs, such glass is called a beam splitter or a half-silvered or semi-transparent mirror (albeit the glass does its job far more precisely than your average windowpane), and as the name suggests, it splits a beam of light into two. Half the energy of the wave is reflected and half is transmitted.

Something funny happens when the incident light is made of just one photon, the smallest indivisible unit of light. It can't be split further into two halves. So the incident photon will be either

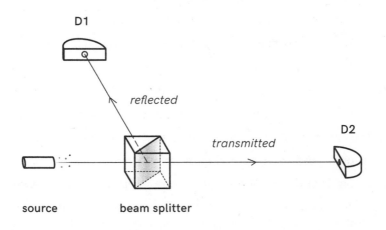

D1

reflected

D2

transmitted

source   beam splitter

transmitted or reflected as one whole. Let's put photon detectors D1 and D2, one at the end of each beam. Since the photon travels undivided, if the photon is reflected, then D1 clicks, and if it's transmitted, D2 clicks. Both detectors *never* click at the same time: the photon does indeed behave as an indivisible unit of energy.

Turns out that if you send lots and lots of photons, one at a time, into the beam splitter, then on average, half the time D1 will click, and half the time D2 will click. There's an important observation here that will become more and more relevant: one can never predict with certainty what a given photon will do. For each photon, we can assign only a probability for the outcome—it'll either go to D1 with probability of 0.5 or to D2 with a probability of 0.5.

The immediate reaction to this from our classically attuned minds is to say, well, the fact that we can't predict what the photon will do must have something to do with our lack of knowledge about the complete state of the photon. It was a similar argument that initially prompted Einstein to think that there must be some hidden variables.

For example, when we flip a coin, we assign a probability of one-half to the outcome that it'll land heads, and one-half that it'll come up tails. But the reason we cannot predict with certainty whether a coin will come up heads or tails is because we don't know everything there is to know about the coin's initial conditions (the angle at which it's flipped, the initial velocity, etc.); if we had full knowledge of the initial conditions, knowledge that could be encoded with additional variables, we could in principle predict the outcome.

Could something similar be at work with the photons? What if there were some attributes of the photon that weren't being captured in the mathematical formalism? And if we knew the values of these hidden variables, then could we not predict with certainty what each photon would do?

Putting aside concerns about hidden variables for now, let's make the experimental setup a bit more interesting. Let's put fully reflecting mirrors in the paths of the beams so that the beams are turned at right angles, toward each other. D1 and D2 are still at the ends of these beams. What can we expect if we send tens of thousands of photons, one at a time, into the apparatus?

Well, nothing different. All we've done is increase the distance traveled by the photons, but nothing else has changed. So D1 still clicks half the time, and D2 half the time.

To complicate things further, let's add a second beam splitter, exactly at the point where the paths of the two beams cross. Now, a photon reaching the second beam splitter will be either reflected or transmitted.

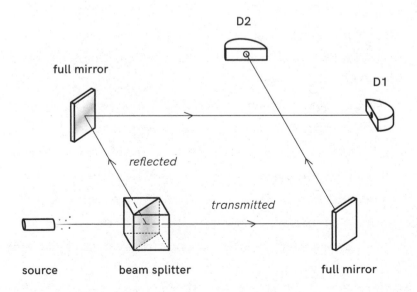

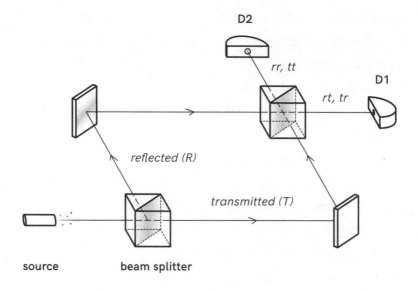

Based on what we know, we can analyze what's expected of each photon that enters the apparatus. A photon that's reflected by the first beam splitter and then transmitted by the second beam splitter ends up at D1. Let's call this photon *rt*. The photon that's transmitted first and reflected next also arrives at D1. This photon is *tr*. So photons that take the paths *rt* and *tr* reach D1. Similarly, *rr* and *tt* end up at D2.

What happens then, if we send 10,000 photons, one at a time, into the apparatus? We know from previous experiments that half of the photons would be sent one way by the first beam splitter and half the other way (speaking always, on average). So 5,000 photons would take path R and 5,000 would take path T (the fate of any given photon, however, is still not for us to predict with certainty). At the second beam splitter, the 5,000 that took path R should be split in half again; 2,500 should go to D1 and 2,500 should go to D2. The same analysis holds for the 5,000 photons that took path T. Total them up and our naive but perfectly logical conclusion is that 5,000 photons should reach D1 and 5,000 should reach D2.

By now, if our minds are getting used to quantum strangeness, we shouldn't be surprised to learn that that's not what happens. Adding the second beam splitter has profound implications. Before the second beam splitter was in place, it was clear when D1 clicked that the photon had come through path R, and if D2 clicked, the photon came via T. Now, with the two beam splitters in place, a photon reaching D1 could have taken either path *rt* or *tr*, and a photon reaching D2 could have taken either path *rr* or *tt*. If you detect a photon at D1, there's no way to tell which path it took. The same goes for a photon detected at D2. The two paths, for any given photon, have become indistinguishable. This is exactly what happens in a double slit. Once a photon lands on the far screen, there's no way to tell which slit the photon came through. If nothing about the experimental setup makes the photon take one *or* the other path, the mathematical formalism says that it takes both paths, in a manner of speaking. Now that we know as much, can we figure out what happens to the 10,000 photons?

The clue is in the name of the apparatus sketched above. It's called a Mach-Zehnder interferometer, after Ludwig Mach (son of the physicist Ernst Mach) and Ludwig Zehnder. In 1892, Mach improved an instrument that Zehnder had designed a year earlier for optics experiments (they weren't thinking of single-particle quantum mechanics at the time). The Mach-Zehnder interferometer is a special case of a double-slit experiment. Light can take one of two paths (analogous to going through one slit or the other) and then interfere when the paths recombine in the second beam splitter (or at the location of the screen in the case of the classic double slit). Conceptually, anything you can do with a classic double slit, you can do with a Mach-Zehnder interferometer: modern experimental physicists, when they say they are performing a double-slit experiment, are in all likelihood using this interferometer. It's an experimentalist's delight. "It's just smarter," Aspect told me.

Why interference happens, especially when we send one particle at a time into the apparatus, is a curious phenomenon. It's exactly what Aspect and Grangier studied when they built such an interferometer.

First, his team had to ensure that there was only one photon passing through their experiment. They began by using carefully calibrated lasers to excite atoms of calcium to a higher energy level. Such an excited atom falls back down to its previous state by emitting two photons, the first one a green photon at a wavelength of 551.3 nanometers (one nanometer equals 10^{-9} meters), followed almost immediately by a blue photon at 422.7 nanometers. So there's a green-blue pair, followed by nothing, then another green-blue pair, then nothing, and so on. The pairs are well separated in time. "Good, I'm going to use this separation in time," Aspect recalled thinking.

The green photon heralds the arrival of the blue photon. So the team used the green photon to get their detectors ready for the arrival of the blue photon, which would come within nanoseconds. The crucial thing here is that there will be one and only one blue photon in the apparatus at that point in time. "The probability of having a second blue photon during that time was peanuts," said Aspect. The initial test was to send the blue photon through the first beam splitter, without the second beam splitter in place. So the photon goes to either D1 or D2. Quantum theory says that only one of D1 or D2 should click for a given photon. This part of the experiment was successful. The blue photon always arrives at one or the other detector—and it's clear which path it took. The two detectors *never* click together. The photon always travels as an undivided whole; it behaves like a particle.

It was time to test the photon's wave nature. The team added the second beam splitter. Now the two paths become indistinguishable. And so, just as Thomas Young observed interference with his split sunbeam, Aspect and Grangier saw it too.

But what's the meaning of interference in a Mach-Zehnder setup for a single photon? For a beam of light, we know that constructive interference results in a bright fringe, and destructive interference results in a dark fringe (no light). It turns out that in the interferometer, when the two paths are of equal length and you send in photons one by one, they *all* go to D1, and none go to D2. So D1 represents constructive interference and D2 represents destructive interference. Our earlier naive analysis of the 10,000 photons—that half should reach D1 and the other half should reach D2—is wrong. All 10,000 photons reach D1 and none arrive at D2.

The only way to explain the result is to think of light as a wave. When the wave encounters a beam splitter, the wave splits; half goes into one arm of the interferometer, and half into the other arm. Zooming into the structure of the beam splitter (which has a finite thickness),

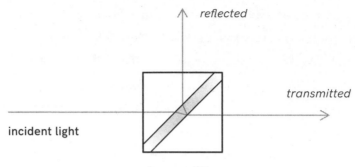

beam splitter

one can see that when a wave is reflected, it travels differently through the glass compared to a wave that's transmitted.

When the wavelength of the light and thickness of the beam splitter are chosen correctly, this can cause the reflected wave to lag behind the transmitted wave by a quarter of a wavelength.

Add in one more reflection at the second beam splitter, and the wave that's reflected twice, *rr*, now lags behind the wave that's transmitted twice, *tt*, by half a wavelength. Both *rr* and *tt* reach D2. So the crest of one wave arrives at D2 at the same time as the trough of the other. That's destructive interference. You get darkness at D2.

The same analysis shows that the *rt* and *tr* paths arrive at D1

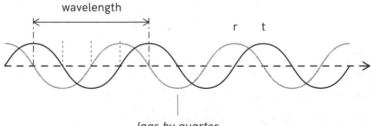

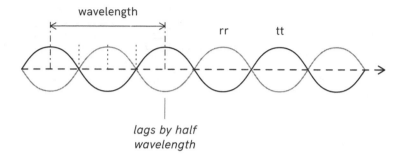

with their waves in sync, the crests of both waves reaching at the same time. The waves are said to be in phase. That's constructive interference. All the light reaches D1.

That's a perfectly fine explanation for light composed of lots of photons and acting like an electromagnetic wave: it's easy to conceptualize half the wave going one way and the other half the other way and then recombining to produce the interference effects. But exactly the same thing happens when you send in single photons. We seem to be getting constructive interference at D1 (all 10,000 photons, hence all the light, reaching D1), and destructive interference at D2 (no photons, hence no light, getting to D2).

This is all very strange. For interference to happen, a wave has to split into two and then recombine. Can that happen to a single photon? When we add the second beam splitter, each photon *seems* to be splitting, going through both arms of the interferometer and recombining. But—and it's worth pondering this to the point of pain—we know that a photon cannot split into two, for it's an indivisible unit of light. What then is going through both arms of the interferometer (or through both slits of a double-slit setup)? Digging further reveals exactly why the quantum world is so confounding.

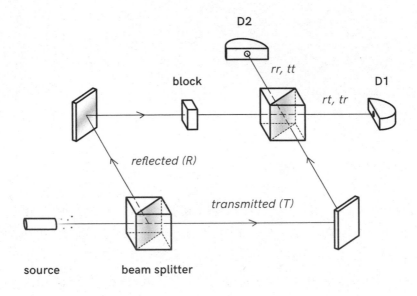

Let's go back to the full Mach-Zehnder setup and make one small change: block one of the arms to prevent any photons from going through. Let's start with blocking arm R. What can we expect?

Two things happen. First, the number of photons that make it to the second beam splitter is halved. By blocking one arm, we are preventing half the photons from getting through to the detectors.

But something else happens that's far more puzzling. *Without* the block in place, we know that all photons leaving the second beam splitter go to D1. However, *with* the block in place, while the number of photons reaching the second beam splitter is reduced to half, half of the ones that do get past the second beam splitter go to D1 and the other half to D2. Repeat the experiment by blocking arm T, and you will get the same results. The block, whether it's in arm R or T, is like a detector: we know *with certainty* that the photons that reach the second beam splitter are traveling through the unblocked arm of the interferometer. There is no indistinguishability anymore, the photons act like particles, and there is no interference, leading to the 50-50 split at D1 and D2.

With the second beam splitter in place and nothing blocking the photon's path, each photon is doing something that is impossible to intuitively understand. It's in a state of *quantum superposition*. As philosopher David Albert of Columbia University writes in his book *Quantum Mechanics and Experience*, the term *superposition* is "just a name for something we don't understand." (And the above analysis is inspired by a similar analysis of a slightly different system in Albert's book.)

The photon is in a superposition of two states, one state in which it goes through one path, and another state in which it goes through another path. But this is not the same as saying it went through both paths, or that it went through only one or the other path, or that it went through neither path.

"The double-slit experiment is usually used to establish that there can be situations in which it makes no sense to ask, say, which slit the particle went through," Albert told me when we met at his home in New York City. "There fails to be a fact of the matter about which slit the particle went through. Asking which slit the particle went through is [like] asking about the marital status of the number five or the weight in grams of Catholicism. This is something philosophers call a category mistake."

Nonetheless, there's interference. What is it that interferes? Interference has to do with quantum superposition, and the story of how quantum objects end up in, and fall out of, superposition is "the *most* unsettling story perhaps, to have emerged from any of the physical sciences since the seventeenth century," writes Albert.

Erwin Schrödinger was particularly unsettled by quantum superposition, and so was Einstein, and they came up with some morbid thought experiments to make their point that taking quantum mechanics as it is could result in untenable macroscopic realities. In doing so, they were taking direct aim at some of the key tenets of the Copenhagen interpretation.

Take the Mach-Zehnder interferometer. According to the formalism, before one of the detectors clicks, signaling the arrival of a photon, the photon is in a superposition of two states, of taking one path *and* the other path. The wavefunction of the photon is said to be in a superposition of two states, where one state represents its progress along one path, and the other state its progress along the other path. Using this wavefunction, we can calculate the probability that it'll be found at D1 or at D2, which, incidentally, turns out to be 1 and 0, respectively, when the length of the paths to D1 and D2 are exactly the same. Creating small differences in the lengths of the paths can change these probabilities. According to the standard view, the photon has no definite position—the wavefunction is spread out—until there's a measurement. The measurement at D1 or D2 causes the wavefunction to *collapse* to one definite value: the photon shows up at one of the detectors.

The onus of collapsing the wavefunction falls on the measurement device, which is assumed to be some macroscopic, classical apparatus. But the Copenhagen interpretation does not really define the exact meaning of a measurement. How big does the measure-ment device have to be to count as classical? Where's the boundary between the quantum and the classical? Such questions lead to the so-called measurement problem.

To appreciate just how deep this problem runs, let's assume the measuring devices are also something quantum mechanical—say they are themselves particles that interact with the photon and somehow change state to record the arrival of the photon. If one were to simply follow the mathematics, something weird happens. The wavefunction of the photon evolves until it is in a superposition of having gone through both paths, which recombine at the second beam splitter. Then it interacts with the measurement device, which is itself a quantum mechanical particle. The whole system then ends up in a superposition, mathematically speaking, of a state in which the photon has reached the measuring particle D1 and D1 has changed state *and* the photon has reached the measuring par-ticle D2 and D2 has changed state. To collapse the wavefunction of the entire system so that we can figure out exactly where the photon went, we'll have to do yet another measurement, with some osten-sibly classical apparatus, to determine the state of D1 and D2.

What if every piece of apparatus we choose obeys the laws of quantum mechanics, such that the entire setup always remains in superposition, and its total wavefunction never collapses? Will it eventually require a conscious human to cause a collapse, to see where the photon went?

The Copenhagen interpretation, while it does not invoke the need for human consciousness, nonetheless demands a classical measurement. The corollary is that the quantum state of a system is adequately and indeed completely captured by the wavefunction, and since the wavefunction lets you calculate only the probability of finding the system in some state, and does not correspond to, say, where the photon actually is, reality does not exist in any meaningful sense independent of measurement with a classical apparatus.

Einstein and Schrödinger were both deeply disturbed by such anti-realist ideas. Einstein pointed out his concerns in a letter to Schrödinger. In a thought experiment, Einstein imagined some gunpowder that can spontaneously combust because of quantum mechanical goings-on and has a certain probability of exploding within a year, and a certain probability of not exploding. The system starts off with a well-defined wavefunction, meaning it is in a definite state. But because it's a quantum system, the wavefunction evolves according to the Schrödinger equation, into something that eventually puts the gunpowder in a quantum superposition of having exploded and not having exploded. Of course, macroscopically, in our understanding of the world at large, this is absurd. The gunpowder should either explode or not, whether we look at it or not. Einstein wrote to Schrödinger, saying, "Through no art of interpretation can this ψ-function [wavefunction] be turned into an adequate description of a real state of affairs; [for] in reality there is just no intermediary between exploded and not-exploded."

Schrödinger ended up refining this thought experiment, making the seeming absurdity starker. He wrote to Einstein, saying, "I am long past the stage where I thought that one can consider the ψ-function as somehow a direct description of reality. In a lengthy essay that I have just written I give an example that is very similar to your exploding powder keg." He then went on to describe the experiment, an elaborate version of which appeared in his published paper.

"A cat is shut up in a steel chamber, together with the following diabolical apparatus (which one must keep out of the direct clutches of the cat): in a Geiger tube there is a tiny mass of radioactive substance, so little that in the course of an hour *perhaps* one atom of it disintegrates, but also with equal probability not even one; if it does happen, the [Geiger] counter responds and through a relay activates a hammer that shatters a little flask of prussic acid. If one has left this entire system to itself for an hour, then one will say to himself that the cat is still living, if in that time no atom has disintegrated. The first atomic disintegration would have poisoned it. The ψ-function of the entire system would express this situation by having the living and the dead cat mixed or smeared out (pardon the expression) in equal parts."

But surely the cat is either alive or dead, not in some weird mixture of both? At least that's our classical intuition. Not so says the standard view of quantum mechanics. The paradox arises because in the Copenhagen interpretation, the wavefunction of the total system remains in a superposition of cat-dead *and* cat-alive states, until something classical interacts with the system. Say someone opens the steel chamber and takes a look, at which point the wavefunction of the entire system collapses to one or the other. The cat will be found either dead *or* alive.

Quantum mechanics asks us to suspend disbelief and hold on to some counterintuitive notions of reality for long enough to be able to appreciate the bizarreness of the subatomic world. For example, even if Schrödinger's thought experiment with the cat-in-a-steel-box stretches credulity, as it should (it was after all an exercise in trying to demonstrate the possible incompleteness of quantum mechanics), it is also a reminder of something that's true of the quantum world: superposition exists, at least in the standard view.

If it did not, you could not explain the interference pattern seen in a double-slit experiment done with single particles. In Aspect and Grangier's Mach-Zehnder setup, until each photon was detected, it *was* in a superposition of having taken both paths. Mathematically, here's what is happening in the Mach-Zehnder experiment. There is one wavefunction that describes the state of the photon going via one path, and another wavefunction for the state of the photon going via the other path. The final wavefunction is a linear combination of the two wavefunctions, which lets you calculate the probabilities of detection at D1 or D2 ($\psi_{total} = a1.\psi_{D1} + a2.\psi_{D2}$, where the probability of finding the photon at D1 is given by the modulus-squared of a1, and the probability of finding it at D2 is the modulus-squared of a2. The total probability has to add up to unity, so $|a1|^2 + |a2|^2 = 1$. The exact values for a1 and a2 depend on the path lengths of the interferometer, whether they are identical or slightly different).

In the early days of quantum mechanics, it was often said that the interference pattern appears because a particle interferes with itself (to paraphrase Paul Dirac). But that turns out to be a somewhat limited view of what's happening. The more profound realization is that what are really interfering are two different states of the system. In the case of an interferometer with two paths, the two states are the two possible paths for each photon. If there were multiple possible states or paths for the photon, say you had five slits instead of two, then the superposition would involve the photon going through all five slits, and a very different interference pattern would emerge on the far screen.

Whether it's the Mach-Zehnder interferometer with two well-defined possible paths, or an apparatus with five slits, the standard quantum formalism makes it impossible to visualize the path of an individual photon—there are no equations to calculate trajectories. The Copenhagen interpretation of this formalism insists that such trajectories don't exist. In fact, the notion of the path has no meaning, just as the notion of an electron's orbit around the nucleus has no meaning. If realism is the idea that the world objectively exists out there, with well-defined properties, even if we are not privy to them, then the Copenhagen view is anti-realist. In its telling, the only world we can talk of definitively is the one that reveals itself upon measurement; talking about anything else is meaningless.

Aspect, for one, is holding out hope for part of Einstein's dream. "I'm really sitting on the side of Einstein," he told me. "I think there is a real world." Meaning a reality independent of observers, experiments, and experimentalists. For now, Aspect is willing to accept quantum mechanics for what it's saying about the world. "The world is not as simple as one could think. But physicists were smart enough to develop mathematical tools to render an account of what happens."

The impossibility of talking about exactly what's happening to a quantum system—say a photon working its way through a Mach-Zehnder interferometer—within the confines of the Copenhagen interpretation was highlighted by theoretical physicist John Wheeler in the late 1970s and early 1980s, most creatively by his use of the metaphor of the "great smoky dragon." Wheeler imagined, and got someone to sketch for him, a dragon whose head and tail are clearly visible. The tail represents the unambiguous quantum state of the photon that is about to enter an interferometer. The head represents the definite detection of the photon at either detector D1 or D2. But the dragon's body is fuzzy, ambiguous, smoky, hence the name. "What the dragon does or looks like in between [the head and the tail] we have no right to speak [of]," wrote Wheeler and his colleague Warner Miller. The dragon's blurry body, of course, represents the state of the photon as it winds its way through the interferometer (if it does that at all).

Wheeler, like Einstein, loved thought experiments. The one most associated with him bears his name: Wheeler's delayed-choice experiment. It's an experiment that's easy to conceptualize, now that we know of Aspect and Grangier's 1985 experiment done with single photons, using a Mach-Zehnder interferometer. Wheeler, of course, thought of it before anyone had done such experiments.

The delayed-choice experiment brings into stark relief Bohr's complementarity principle. Bohr argued that the wave nature and particle nature of a quantum system are mutually exclusive ways of looking at reality: what you observe depends on the experimental setup, and you cannot have both types of setups in the same experiment. And if you try to do both (as Einstein tried to with his thought experiment with the double slit), then Bohr's claim was that the uncertainty principle ensures that you cannot see the interference pattern (the actual explanation, it'll turn out, is more involved than Bohr understood).

So a quantum system has two faces. But when does it "decide" to show one face or the other? Is that even a legitimate question?

Wheeler took on the question in dramatic fashion. His thought experiment uses the Mach-Zehnder interferometer, in which we observe the photon's particle nature when we don't have the second beam splitter in place (D1 and D2 each click half the time) and we observe the photon's wave nature when the second beam splitter is in place (there's interference, and D1 clicks all the time, and D2 never does).

Here's what Wheeler proposed. What if we delay the choice of whether or not to put in place the second beam splitter until after the photon has gone past the first beam splitter and thus entered the interferometer? How does the photon "know" what to do inside the interferometer? At the moment it encounters the first beam splitter, let's say there is no second beam splitter to measure the wave nature. From all we know thus far, the photon should go through either one arm or the other, as a particle. After the photon is on its way to either D1 or D2, we insert the second beam splitter and make the two paths indistinguishable. The experimental setup is now looking for the photon's wave nature. What's the photon to do? Suddenly decide that it has to go into a superposition of taking both paths and display interference?

You can also do the opposite. Let the photon enter the interferometer with the second beam splitter in place—so now it *is* in a superposition of going through both paths, which means eventually it'll end up at D1, not at D2. But let's take out the second beam splitter just before the photon encounters it. If we continue thinking in terms of one path taken or both paths taken, the photon has to somehow do extreme calisthenics and appear to take only one of two paths, and thus reveal its particle nature, and end up at either

D1 or D2. It's *as if* the photon is going back in time and undoing what it had done. As Wheeler wrote: "One decides whether the photon 'shall have come by one route or by both routes' after it has *'already done* its travel.'"

The italicization of, or the use of scare quotes around, words and phrases like *seems*, *knows*, *as if* is deliberate—they highlight the fact that our classical notions and the language we use to talk about them fail us when dealing with the quantum world, at least when we limit ourselves to the standard formalism and the Copenhagen interpretation.

When Aspect did the single photon double-slit experiment in 1985, he was aware of Wheeler's thought experiment, and he was aware that his Mach-Zehnder setup had some of the necessary ingredients for testing Wheeler's ideas. Still, "at that time I was not dreaming of doing the experiment," Aspect told me.

The reason being that the delayed-choice experiment is technically far more challenging. In Aspect's initial experiment, each arm of the Mach-Zehnder interferometer was about 6 meters long, which is the distance from the first beam splitter to one of the detectors. A photon takes only about 20 nanoseconds to cover that distance. To trick the photon after it had entered the interferometer, Aspect's team would have had hardly any time to either insert or remove the second beam splitter. It seemed impossible.

Why not just increase the length of the interferometer? Say to 50 meters? That would give one about 165 nanoseconds to do the dirty deed. It's not eternity, but not an impossibly small interval of time either.

"I'm going to teach you something about optics," said Aspect, and proceeded to explain why going long was not an option in 1985. Their source of single photons then was not point-like. It was as if the photons were coming out not from a pinhole but from an opening with a larger diameter. In optics experiments, you often need lenses to corral photons toward your detectors. And if the source is not point-like, the light can diverge, making it necessary to build bigger and bigger lenses, which get prohibitively expensive and technologically infeasible. "Six meters was already a problem, because my source was not exactly point-like, but I could solve it," said Aspect. "But 50 meters was out of the question." That would have required lenses several meters in diameter.

Aspect had to wait twenty years before the technology caught up to where he could do Wheeler's delayed-choice experiment just as Wheeler had intended. Other teams, meanwhile, had done versions of the experiment, but Aspect was after its essence and did not want to leave anything to interpretation. "It was clear for me in 2005 that the technology had reached a point where you can [do] an experiment which is very close to the ideal scheme of Wheeler," he said.

By 2005, Aspect had a source of single photons that was much more point-like, and he was able to build an interferometer with arm lengths of 48 meters—enough time to insert or remove the second beam splitter after the photon had passed the first beam splitter.

And what they observed was that there was no fooling the photon. If the second beam splitter was not there, it behaved like a particle, otherwise it acted like a wave. It did not matter *when* the second beam splitter was inserted.

Recall that the initial arguments between Bohr and Einstein as to why one cannot observe the wave nature and particle nature simultaneously had to do with Bohr's assertion that the act of observation somehow disturbed the apparatus, smearing out the interference pattern. Complementarity was the outcome of the uncertainty principle.

But the delayed-choice experiment demonstrates that complementarity is a deeper principle, deeper than probably Bohr realized. In Aspect's 2005 experiment (the results were published in February 2007), the photon as it goes past the first beam splitter is still too far away from the location of the second beam splitter to be influenced by the decision being made at the distant location. In the language of special relativity, these two events are space-like separated, so there is no question of any disturbance due to measurement; nothing that is being done near the output stage can influence the photon. And yet, the photon shows only one face or the other.

"Bohr's statement that it is the measurement that determines what you observe etc.... should not be taken in a too naive [manner]. It is more subtle than that," said Aspect.

Just how subtle would become abundantly clear with an even more audacious version of the delayed-choice experiment: the so-called delayed-choice *quantum eraser* experiment. This is to distinguish it from Wheeler's original idea, in which the second beam splitter is a classical, macroscopic device. It's either there *or* not there. What if the experiment not only delayed the choice of whether to look for the particle nature or the wave nature of photons but allowed for that choice to be erased? What would the photon do?

To understand the delayed-choice quantum eraser experiment

involves going back, yet again, to Einstein's objections to quantum physics. For all his work on the special and general theories of relativity and the photoelectric effect, Einstein's most cited paper is one he wrote in 1935, identifying a weird property of quantum systems (which he would later refer to as "spooky action at a distance"). It's a property that Schrödinger also identified in the same year, and he called it entanglement. If quantum superposition exhibited by single particles was the first mystery thrown up by quantum mechanics, then entanglement, which involves two or more particles, was something even more profound, and for Einstein, fundamentally more disturbing. Aspect calls the developments that followed— including experimental variations of the double slit that outdid even Wheeler's thought experiments—the second quantum revolution. "[It] has to do with realizing that entanglement is dramatically different," he said.

4

FROM SACRED TEXTS

Revelations about Spooky
Action at a Distance

Nonlocality forces us to extend the conceptual toolbox we
use to talk about nature's inner workings.

—Nicolas Gisin

Tim Maudlin can vividly recall the moment the strangeness hit
him. It was more than three decades ago. He was in his senior
year of college, when he came across a *Scientific American* article
titled "The Quantum Theory and Reality" by theoretical physicist
Bernard d'Espagnat. It had a rather long and unwieldly subtitle—
"The Doctrine That the World Is Made Up of Objects Whose Existence Is Independent of Human Consciousness Turns Out to Be in
Conflict with Quantum Mechanics and with Facts Established by
Experiment"—to go with some fifteen-odd pages of text, equations,
and illustrations. Pretty heavy going. Maudlin read it thoroughly
enough to be floored by the implications. "My roommates said later

that they knew something was strange, because I just kept . . . holding this magazine and pacing around in circles in the room," said Maudlin, a philosopher of science at New York University. We were sitting in the sparingly furnished living room of his New York apartment. A framed print of artwork by Croatian artist Danino Bozic graced one wall—as it does the cover of one of Maudlin's books, *The Metaphysics within Physics*. Two thin, tall wooden figures—carvings made by the Nyamwezi people of Tanzania—stood in one corner. In that tastefully austere setting, Maudlin cast his mind back to the d'Espagnat article. He said he'd now take issue with some of the things that d'Espagnat wrote, but reading it then as a student, "it was clear enough that you could see something very strange was going on [with quantum mechanics], that it was a sharp enough result that you couldn't get out of it."

The *Scientific American* article had a detailed exposition of John Bell's 1964 paper, the very same paper that inspired Alain Aspect to embark on his experiment to settle a debate between Einstein and Bohr. In the article, d'Espagnat argued that Einstein's ideas (enshrined in his theories of relativity) and quantum mechanics were at odds, and that Aspect's experiment to test Bell's theorem, which had yet to be done, would settle the issue. Einstein became aware of this tension between his theories and quantum mechanics well before almost anyone else.

To make his case at the 1927 Solvay Conference, Einstein took the case of a particle going through a hole in a screen. With characteristic humility, he "first apologized for not having gone deeply into quantum mechanics." Then, with characteristic insight, he gave an astute analysis of what was puzzling him. According to the formalism, the particle's wavefunction goes through the hole, diffracts, and spreads out semi-spherically, and one can calculate the probability of finding the particle at any one location on the surface of this spreading hemisphere. Now, let's say a detector at some distance from the hole detects the particle. This is the same as saying that the spread-out wavefunction collapses upon measurement. If one interprets the wavefunction as a complete description of the state of the particle and as a description of what's actually happening (and not merely a statement about our state of knowledge, or lack thereof, of reality), then it seems that the particle, which was itself spread out, gets localized. If so, Einstein made the point that this localization, or the unequivocal appearance of the particle at one location, is happening simultaneously with the indisputable disappearance of the particle from all other locations. It's a violation of the principle of locality, which says that if something is happening in one region of spacetime, it cannot influence something else happening in another region of spacetime any faster than the speed of light. The collapse of the wavefunction, in this way of thinking about it, is instantaneous and patently nonlocal. Even this early in the history of quantum mechanics, Einstein was aware that this seeming nonlocality, which implied simultaneity of actions, was in conflict with his own theory of special relativity.

But a more seminal analysis was to come from Einstein and two colleagues. To Einstein's chagrin, the world learned of this not through the usual channels of scientific discourse but via an almost tabloid-like report in *The New York Times*.

EINSTEIN ATTACKS QUANTUM THEORY screamed the headline on May 4, 1935. According to *The New York Times*, Einstein and his two younger collaborators, Boris Podolsky and Nathan Rosen, had shown that quantum mechanics wasn't complete and that it needed augmenting.

Einstein found out that Podolsky had leaked information about their upcoming paper some two weeks in advance of its publication in *Physical Review*, and Einstein complained to the *Times*, saying, "I deprecate advance publication of any announcement in regard to ... [scientific] matters in the secular press." (Decades later, physicist David Mermin would quip, "If *The New York Times* is the secular press, it follows that the sacred text is the *Physical Review*.")

Nonetheless, the elegant four-page-long paper, published on May 15, 1935, was a slowly unfolding seismic event whose aftershocks continue to shake up quantum physics to this day. Known as the EPR paper in the literature (for Einstein-Podolsky-Rosen), it had its roots in a teatime conversation between Einstein and Rosen about the quantum state of two particles after they have interacted with each other. It turns out that post-interaction, there is no separate wavefunction to describe each particle; rather, they are described with one, joint wavefunction. The particles are said to be entangled. Einstein figured that entangled particles would allow him to strengthen his argument about the incompleteness of quantum theory. Despite his seeming defeat at the hands of Bohr at the Fifth Solvay Conference, giving the round to the Copenhagen interpretation, Einstein was far from done debating Bohr.

Central to the EPR argument is the assumption that nature is local. It seems a pretty intuitive idea, but it took Einstein's theory of general relativity to bring locality into sharp focus. Before Einstein, Newton had obfuscated locality by suggesting that gravitational influences were instantaneous. In Newtonian gravity, if the sun were to somehow disappear, the Earth would be immediately influenced by the changes in the gravitational field. Einstein's general relativity showed that gravity is the outcome of the warping of spacetime by the presence of matter (the way a heavy ball placed on a taut sheet of rubber dents the sheet), and any changes to the curvature of spacetime caused by matter can propagate only at the speed of light. So it will take Earth about eight minutes to notice the absence of the sun's gravitational pull, were it to disappear. Locality is essential to Einstein's relativity.

Besides locality, Einstein had long held dear the idea of "realism" and it showed up in the EPR paper, which begins with these words: "Any serious consideration of a physical theory must take into account the distinction between the objective reality, which is independent of any theory, and the physical concepts with which the theory operates. These concepts are intended to correspond with the objective reality, and by means of these concepts we picture this reality to ourselves."

For Einstein, the real world exists independent of our observations.

Realism can be sharpened to a statement about our physical theories, to argue that variables in the theories correspond to actual physical reality. Completeness of a theory, in this regard, depends on it having enough relevant variables to capture all of physical reality (for example, variables for a particle's position and momentum, which would allow us to calculate its trajectory, if a particle has one, that is).

And indeed, one of the arguments EPR used to make their point that quantum mechanics is incomplete involved a somewhat complicated thought experiment that required measuring the position and momentum of entangled particles. Sixteen years later, in 1951, physicist David Bohm would illustrate the EPR argument using a simpler thought experiment. With hindsight, it's easier to understand the issue using Bohm's clearer example.

Imagine a particle with zero spin that decays into two identical particles that move away from each other. The conservation of angular momentum dictates that the particles must be spinning in opposite directions, so that the total spin still adds up to zero. Assume that the particles, A and B, are moving away from each other along the X-axis (the left-right direction on this page). Quantum mechanics says that the two particles are entangled in their spin.

Let's first take particle A. If you were to measure the spin of the particle along the X direction, you can predict only the probability of the outcome. It'll be either UP or DOWN. The same holds true if you were to measure the spin in the Y direction (the up-down direction on this page) or the Z direction (an axis going in and out of this page), or in any arbitrary direction.

This is also the case if you were to measure *only* the spin of particle B in any arbitrarily chosen direction. There's absolutely no way of predicting with certainty the result of the measurement.

Now comes the part that bothered Einstein. A and B are entangled. So, if you were to measure A's spin in the X direction, and find it to be UP, the quantum formalism tells you with *absolute certainty* that B will have a spin of DOWN in the X direction. You don't have to measure B to know it to be so, but if you do the measurement, it *will be* so. If A is measured first, then the fate of B's spin is sealed, and vice versa, as long as you measure the spins of both particles in the same direction, say along the X-axis. It's *as if* what we did at the location of particle A instantly influenced particle B—a form of apparent nonlocality.

The EPR argument explicitly assumed locality, making such influences impossible. With this assumption, EPR could make an argument for the existence of the reality of the spin of particle B: "*If, without in any way disturbing a system, we can predict with certainty (i.e., with probability equal to unity) the value of a physical quantity, then there exists an element of physical reality corresponding to this physical quantity*" [italics in original].

That's exactly what the measurement of particle A achieves. If the world is local, then what we do at the location of particle A cannot disturb particle B. Yet we can now instantly predict with certainty the value of the spin of particle B, no matter how far away B is from A. So particle B must have had that value before the measurement at A. And if a particle has a property whose value does not depend on a measurement, then it's possible to have a variable in the theory that captures that property. You can see where this is going. All quantum mechanics has is a wavefunction that tells you the probability of outcomes of measurements; it does not have such hidden variables (say, for the position of a particle; the fact that a variable for the position of a particle would be called hidden was "a piece of historical silliness," wrote Bell).

The EPR paper triumphantly concluded that "the wavefunction does not provide a complete description of physical reality." They did not address what a complete theory would look like but said, "We believe, however, that such a theory is possible."

Any such theory that augments the wavefunction with additional variables came to be called a hidden variable theory. The irony of the EPR paper was that a few years earlier, in 1932, John von Neumann had ostensibly proved that any theory that could replicate the experimental successes of quantum theory could not have hidden variables. Those who swore by the Copenhagen take on the quantum world were only too happy to accept von Neumann's proof.

So much so that one philosopher's protestations were lost to the world. Grete Hermann was a German philosopher, and "the first and only female doctoral student" of Emmy Noether, a formidable mathematician whose work underpins much of modern physics. Hermann straddled the worlds of philosophy and mathematics with ease. In 1935, she published a paper in a German journal in which she showed that von Neumann's proof was incorrect. "A thorough examination of the proof of von Neumann reveals . . . that in his argumentation he makes an assumption which is equivalent to the statement he wants to prove," she wrote. "Therefore, the proof is circular."

Even Einstein, around 1938, is reported to have said of the assumption Hermann identified: "Why should we believe in that?" Einstein, of course, was increasingly being thought of as a curmudgeonly old man who was holding on to his precious ideas of realism, locality, and at times even determinism. Though, to be fair to Einstein, the lack of determinism did not overly bother him. The overused quote of his in popular culture, that "God does not play dice with the world," misrepresents his stand on the issue. He certainly did, during the early 1920s, express concerns about the indeterminate nature of the quantum world, saying that he found the idea "intolerable," and if it were true, he "would rather be a cobbler, or even an employee of a gaming house, than a physicist." But as quantum mechanics matured, Einstein backed off from his disavowal of indeterminism. It was an aspect of reality he was willing to accept. Not so with anti-realism and nonlocality. In any case, it was easy for the younger crowd to be dismissive of Einstein's views as he grew older.

The reasons why Hermann's work never gained widespread attention are less clear. Publishing in an obscure German philosophy journal didn't help. But that's not the entire explanation, since Heisenberg and his colleagues were aware of her work. Maybe, swayed by their own ideas, they overlooked the implications of Hermann's claims. Political affiliations supposedly played a part. Or maybe it was the sexism of the time, argues philosopher Patricia Shipley, but adds, "If that had something to do with it, I don't think it was the primary reason, it could have been a secondary reason."

It was David Bohm who, in 1952, a year after he reworked Einstein's EPR argument into a simpler thought experiment, implicitly undermined von Neumann's proof by constructing a hidden variable theory. Decades later, John Bell would say, "In 1952, I saw the impossible done," referring to Bohm's formulation that went against von Neumann's impossibility proof.

In an interview with *Omni* magazine in 1988, Bell was scathing: "The von Neumann proof, if you actually come to grips with it, falls apart in your hands! There is *nothing* to it. It's not just flawed, it's *silly*. If you look at the assumptions made, it does not hold up for a moment. It's the work of a mathematician, and he makes assumptions that have a mathematical symmetry to them. When you

translate them into terms of physical disposition, they're nonsense . . .
The proof of von Neumann is not merely false but *foolish*."

Unimpressed by von Neumann's proof, but inspired by Bohm
and Einstein, Bell saw a way to turn the EPR argument into a test of
quantum mechanics. The result was his 1964 paper, with his eponymous theorem.

Here's one description of an experiment based on Bell's theorem. It's a slight variation of Bohm's EPR thought experiment but is
closer in spirit to what experimentalists like Aspect actually do. It
involves photons of light and a property called polarization.

We saw earlier that light is an electromagnetic wave, so it's got
an oscillating electric field and an oscillating magnetic field. Polarization has to do with the plane in which the electric field is vibrating relative to the direction in which the light is traveling. For
example, if light is moving along the X direction (again, left to right

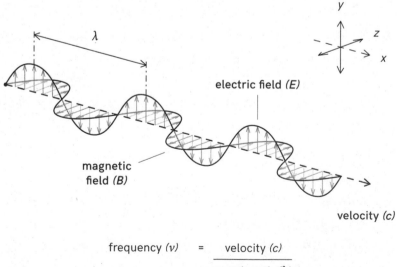

on the page), then the electric field could be oscillating in the X-Y plane (up-down). If so, the light is said to be vertically polarized. If the electric field is vibrating in the X-Z plane (in-out), the light is horizontally polarized. But those are not the only allowed values: the electric field can vibrate in any plane that's at an arbitrary angle to the vertical—this is the so-called angle of polarization. Also, the angle of polarization can hold steady as the light travels, or it can keep changing. Individual photons of light, which are pulses of electromagnetic waves, are also polarized.

Imagine now a source that spits out two photons entangled in their polarization that start moving away from each other toward two observers, Alice and Bob. Alice does one of two measurements on the photon: she checks to see whether the photon is polarized either in the A direction or in the B direction. Quantum mechanics tells us that for each type of measurement, she'll get either a YES or a NO for an answer. Similarly, Bob checks to see if his photon is polarized in one of two directions of his choosing, say, C or D. They repeat this for many, many pairs of photons that come from the source.

Crucially, for each photon pair, Alice and Bob make their measurements independently of each other: neither knows the direction the other is choosing for the measurement.

Now, if the photons that leave the source are not entangled, the outcomes of measurements done by Alice will have no correlation with the outcomes of measurements done by Bob, beyond what's expected by random chance.

But we know that the photons are entangled and quantum mechanics says they are described by the same wavefunction (as far as their polarization is concerned). So, for a given entangled pair of photons, if Alice were to measure her photon's polarization in some direction and get an answer of YES, then we can predict with certainty that if Bob measures his photon in the same direction, he'll get an answer of NO, and vice versa.

Here's where Bell's theorem comes in. For measurements in which the polarization directions used by Alice and Bob are not the same, Bell calculated the amount of correlation that one can expect if Einstein was correct in his assertion that there must be a hidden variable theory that underpins quantum mechanics and also obeys the laws of locality. The correlation is a measure of how many times Alice and Bob would have got contradictory answers.

Bell showed that if Einstein is correct, the correlation has to be less than or equal to a certain amount (hence it's called the Bell inequality test). More specifically, Bell showed that if quantum mechanics is correct and the measurement of a photon's polarization by Alice does instantly influence the state of Bob's photon (and vice versa), then the amount of correlation should exceed that threshold, thus violating the inequality. If so, the quantum world would be manifestly nonlocal.

Soon after Bell published his theorem, experimentalists started testing the inequality. These were not variations of the double-slit experiment, but their findings would have tremendous import for understanding the double slit's essential mystery. Among the forerunners who did such Bell experiments were, most notably, Stuart Freedman and John Clauser at the University of California, Berkeley, Richard Holt and Francis Pipkin at Harvard University, and Edward Fry and Randall Thompson at Texas A&M University. By 1976, a total of seven such experiments had been done, and while two of these experiments disagreed with quantum mechanics (in that they did not violate the Bell inequality), the consensus was that quantum mechanics was correct. The world, at its most fundamental, seemed nonlocal.

It was then that a young Aspect came into the picture. He realized that the ideal experiment as imagined by Bell had yet to be done using single pairs of entangled photons in such a way that the measurements on each pair of photons were space-like separated (so that there was no way that nature could, through some unknown mechanism, let Alice and Bob know of each other's measurement settings any faster than the speed of light). This meant choosing the settings

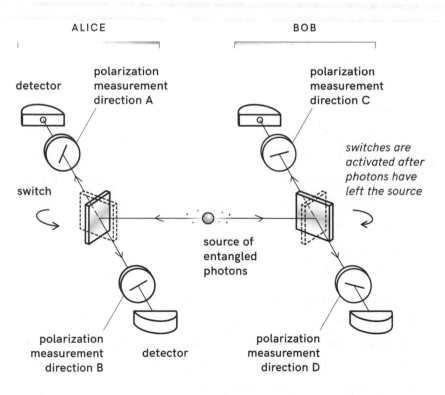

ALICE

BOB

detector

polarization
measurement
direction A

polarization
measurement
direction C

*switches are
activated after
photons have
left the source*

switch

source of
entangled
photons

polarization
measurement
direction B

detector

polarization
measurement
direction D

for the measurement devices—the direction in which to measure the polarization (A or B for Alice, C or D for Bob)—on the fly at either end. The settings, literally, had to be chosen while the photons were in flight from the source to the detectors.

The other challenge was "to build a source of entangled photons which would be able to deliver enough pairs of entangled photons per second," Aspect told me. "At the end, any experiment boils down to the signal-to-noise ratio."

As mentioned earlier, Aspect succeeded in building such a source (it was the technology he'd later use for the single photon double-slit experiment). "It took me five years. By 1980, I had a fantastic source of entangled photons. It was by far the best source of entangled photons in the world," he said. "What Clauser would do in one day, what Fry would do in one hour, I could do in one minute."

With the mass of statistics and the space-like separation between Alice's and Bob's measurements, Aspect was able to show—unequivocally—that the Bell inequality is violated by quantum mechanics. There remained subtle, nitpicky ways in which Alice's measurements could influence Bob's device and vice versa, but those were for the purists. For most physicists, Aspect's experiment had sealed the deal. The experiment made Aspect a star of the lecture circuit (and brought him in touch with Feynman). "It allowed me to propagate the idea that Bell's theorem was really something very important, and that yes, Bell's inequalities are violated, so there is something in entanglement which goes beyond all our ideas of how the world works," Aspect said.

That something is nonlocality. This was a setback for Einstein's hope for a local, realistic hidden variable theory. He never lived to see the results of these experiments, and one can only wonder how he'd react to the growing realization among many followers of standard quantum mechanics that reality is nonlocal. He'd write to Max Born in a letter dated March 3, 1947, "I cannot seriously believe in it [quantum theory] because the theory cannot be reconciled with the idea that physics should represent a reality in time and space, free from *spooky actions at a distance.*" Einstein died in 1955.

Under the gaze of the tall Nyamwezi figures, in his apartment overlooking one of New York's greenest squares, Tim Maudlin explained why entanglement and nonlocality make the double-slit experiment even more intriguing than mere wave-particle duality. With his hands he mimed the wavefunction splitting into two parts, one going through one slit and the other through the second slit. These two parts, as they spread out from each slit, evolve independently and eventually interfere. To calculate the probability of finding the particle at some location away from the two slits, you have to take into account a linear superposition of these two wavefunctions. Say the combined wavefunction hits a photographic plate. The particle appears somewhere on that plate: it gets localized. But at *all* the other locations on the photographic plate where the particle had a nonzero probability of existing, nothing happens. These are simultaneous events and nonlocal.

"How puzzling is that?" said Maudlin.

Do this for particle after particle, and an interference pattern emerges on the photographic plate. The standard analysis of the double-slit experiment usually highlights the appearance of this pattern as emblematic of the mystery of quantum mechanics. In one sense, it undoubtedly is. Each spot that the particle makes on a photographic plate is indicative of both something delocalized—the wavefunction?—going through both slits, and the nonlocal events that result in the particle seemingly appearing at one location on the photographic plate, and simultaneously disappearing from everywhere else.

But to Maudlin, the mystery of the double slit is even more pronounced when one tries to detect which slit the particle goes through. The interference pattern goes away. But why? It's because the system being used to detect the particle as it goes through the double slit becomes *entangled* with the particle. "Schrödinger said that what was really new about quantum mechanics was entanglement," said Maudlin. "And so from that point of view, the really [surprising] quantum mechanical effect is the disappearance of the interference."

The double-slit experiment doesn't merely embody wave-particle duality, the "central mystery," as Feynman said; it incorporates entanglement too. Once physicists began appreciating this, it made possible a new wave of double-slit experiments, each probing deeper into the mysteries of the quantum world. It made possible the delayed-choice quantum eraser experiment.

5

TO ERASE OR
NOT TO ERASE

Mountaintop Experiments Take Us to the Edge

> These experiments are a magnificent affront to our conven-
> tional notions of space and time. Something that takes place
> long after and far away from something else nevertheless is
> vital to our description of that something else. By any
> classical—commonsense—reckoning, that's, well, crazy. Of
> course, that's the point: classical reckoning is the wrong kind
> of reckoning to use in a quantum universe.
>
> —**Brian Greene**

Experimental quantum physicists prefer lab benches and tightly controlled environments. So it's highly unusual that some of the most intriguing experiments in quantum mechanics have been done atop mountains in the Canary Islands, an archipelago just off the coast of northwestern Africa. On a clear day, from the summit of Roque de los Muchachos, the 2,400-meter-high mountain on the small island of La Palma, one can see straight across the blue waters of the Atlantic Ocean to the tops of the volcanic mountains on Tenerife, the archipelago's biggest island, about 144 kilometers away. The experiments, however, have to be done after the sun has set and the moon is still below the horizon, with only the Milky Way spread across the night sky. The foreboding darkness is essential, for the experiments involve sending single photons toward Tenerife's Mount Teide, in the shadow of which a telescope has its sights trained on the photon source at La Palma.

The driving force behind these experiments is the Austrian physicist Anton Zeilinger. He and Alain Aspect are compatriots. Both forged their reputations as clear-thinking experimentalists at about the same time (and were recognized for their efforts in 2010, when they won, along with John Clauser, the Wolf Prize in Physics). But Zeilinger and Aspect couldn't be further apart when it comes to interpreting quantum physics. Aspect, as we saw in the previous chapter, leans toward being a realist in the mold of Einstein. Zeilinger takes after Bohr.

"All quantum mechanics gives us is probability distributions for possible measurement results," he told me. Their tests of Bell's inequality (Zeilinger's team did more sophisticated versions of Aspect's pioneering experiment) have shown that there is no local hidden reality that quantum mechanics is failing to capture. The probabilities one observes, in the Copenhagen view, don't seem to be the outcome of lack of information the way probabilities of outcomes in classical physics of, say, a throw of dice are the result of incomplete information. The Copenhagen followers regard probabilities as intrinsic to quantum mechanics.

"And that is amazing," said Zeilinger when I met him at his office on Boltzmanngasse in Vienna, Austria, a few days after I met Aspect in Paris. Zeilinger's office is a short walk from the Donaukanal, a waterway of the river Danube. The region reeks of history. There are, of course, the street names: Boltzmanngasse for Ludwig Boltzmann, a stalwart of late-nineteenth-century physics and a key figure behind the development of the kinetic theory of gases and statistical mechanics, both of which leaned heavily on probability theory. A few doors away from Zeilinger's building is the Erwin Schrödinger International Institute for Mathematics and Physics, which, before it moved to Boltzmanngasse in 1996, was housed in Schrödinger's home a few hundred meters away on Pasteurgasse, a street named after Louis Pasteur. If the influence of science, especially physics and mathematics, is overwhelming, there are the Sigmund Freud and the Strauss museums, each about a ten-minute walk away.

And so it was that in a building on a street named after a man who put probabilities into classical physics, Zeilinger expressed wonderment at probability's role in quantum physics. "How can that be? How can we just have probability distributions and nothing behind it?"

Then in the very next breath, he said, "The probabilities are the reality we have. There is nothing behind it. The probability is not about a hidden reality . . . full stop." He added he's probably a "non-realist" but said he hates labels. "They are silly categories," he said.

But despite going against Einstein's views, Zeilinger professed enormous respect for his impact on quantum physics. "Sometimes people belittle Einstein's contribution, which is wrong," said Zeilinger.

Einstein, more often than not, is recalled as raising concerns about aspects of quantum mechanics that did not make, well, classical sense. But Einstein did more than that. "He pointed the finger at these things not because as some people say he did not understand quantum mechanics," said Zeilinger, but because he understood it very well. Zeilinger mused about what Einstein would have made of their experiments. "I'd give a lot to hear his comments about this situation," he said. With a light laugh, he said he'd ask Einstein, "You know our results, what do you say?"

At the least, Einstein would likely have been enthralled by the experiments done on mountaintops, given his own penchant for hiking in the Swiss Alps (in 1913, he crossed the nearly 1,800-meter-high Maloja Pass on foot, with Marie Curie and her daughters for company). One mountaintop experiment, a particularly intricate and involved variant of the double-slit experiment done by Zeilinger and his team, combined the two elements of quantum mechanics that made Einstein insist on the theory's incompleteness: wave-particle duality and nonlocality. The origins of this line of inquiry lie in a thought experiment dreamt up by a physicist who came to be called the "Quantum Cowboy," for his pioneering research on the nature of reality and beef cattle production.

During the American Civil War, a confederate officer named Robert P. Salter cultivated cotton on a farm that lies midway between Houston and Dallas. He bought guns with the cotton he grew. Today, Marlan Scully studies sustainable farming on parts of that historic farm. "The mystery is not that I'm interested in farming, but that I'm interested in quantum physics" has been Scully's response to questions about why a quantum physicist took up farming. Scully grew up in rural Wyoming and married into a farming family.

He went to Yale to do his graduate studies, where he pursued the great experimentalist Willis Lamb nearly every day. "A dumb kid from Wyoming, I didn't know that the Nobel Prize physicist at Yale wasn't there for me." Lamb always obliged with his time. After his PhD, Scully continued as an instructor at Yale. Within two years he moved to MIT, and then soon after to the University of Arizona. A decade later he moved to the University of New Mexico, and when he was there, he collaborated with Kai Drühl, a postdoc based in Munich, Germany, to come up with one of the most famous thought experiments in quantum physics: the quantum eraser.

The "quantum eraser is qualitatively, conceptually, intellectu-ally, much deeper than the Young [double slit] experiment," Scully told me during a phone conversation. Still, at its core, it is yet another type of double-slit experiment, albeit a very sophisticated one.

Scully and Drühl targeted a key aspect of the debate between Einstein and Bohr: whether or not experiments themselves disturb quantum systems in ways that enforce complementarity. In the early days of their deliberations, Bohr had argued that the uncertainty principle would prevent us from seeing the wave nature and the particle nature of reality simultaneously. These were complementary aspects that were forever separated by the clumsiness of our classical measurements, and the uncertainty principle was the enforcer. But as Aspect showed with his implementation of Wheeler's delayed-choice experiment, even when you could not point the finger at disturbances caused by the measuring apparatus and hence at the uncertainty principle, complementarity still reigned. It was a deeper principle than anyone had realized. Scully and Drühl pushed the argument much further.

They imagined collecting information about which slit a particle goes through without disturbing the particle. The particle continues to do what it normally does, and yet somehow, it leaves behind information about the path it takes through the double slit. According to quantum mechanics, the mere presence of such information should destroy the interference pattern. As if that isn't surprising enough, Scully and Drühl then asked a deeper question: What if this information is erased? Will the interference pattern come back?

With their thought experiment, the duo was trying to refine the notion of measurement in quantum physics. In the 1930s, John von Neumann developed the rigorous mathematical formalism for quantum mechanics (in the very book in which he supposedly proved that there can be no hidden variable theories). This formalism emerged from axioms that gave measurements center stage: measurements caused a wavefunction to collapse. But there was no precise definition of what constitutes a measurement. Bohr, for example, merely divided up the world into the big and small, and measurement apparatuses were "big," while the things they were measuring were "small." The boundary between the classical and the quantum was entirely unclear—nothing in the formalism suggested where such a boundary might lie.

Yet, the practical use of the theory implied such a boundary. A quantum system, described by its wavefunction, evolves according to the Schrödinger equation and then suddenly, upon measurement, the wavefunction collapses. The process of collapse does not follow the same laws as the ones governing the evolution of the wavefunction. In fact, there is no law, so to speak, that governs collapse. It's something ad hoc, attributed to measurement. So a particle that, until the measurement, was in a superposition of multiple states is reduced to being in just one of the many possible states. What is it that determines when and how this collapse happens?

This question was pushed to, some would say, its logical conclusion by Eugene Wigner, a Nobel Prize–winning physicist and von Neumann's contemporary at Princeton University in the early to mid-1930s. Wigner, after a careful analysis of von Neumann's formalism, concluded that the laws of quantum mechanics did not draw a line between the quantum and the classical. Everything—the quantum system, the measuring apparatus, everything—should evolve according to the same laws. The only thing, he reasoned, that could be responsible for the collapse of the wavefunction was consciousness. The act of perception by a conscious observer, Wigner argued, is the nail in the coffin for the wavefunction. In 1961, he wrote: "When the province of physical theory was extended to encompass microscopic phenomena, through the creation of quantum mechanics, the concept of consciousness came to the fore again: it was not possible to formulate the laws of quantum mechanics in a fully consistent way without reference to the consciousness." But by 1970, Wigner changed his mind, doubting his own claims of consciousness playing a role in causing collapse.

Very few physicists today put stock in Wigner's ideas. Scully and Drühl too weren't concerned about consciousness and its role; they wanted a sharper understanding of measurement and the nature of collapse. They asked whether measurement could itself be something quantum mechanical. If so, the measurement device would also evolve according to the Schrödinger equation. Its wavefunction would not collapse, and so could be made to reverse its evolution in a manner that undid the measurement. "We propose and analyze an experiment such that the presence of information accessible to an observer and the subsequent 'eraser' of this information should qualitatively change the outcome of our experiment."

They designed their thought experiment to show that one could in principle acquire which-way information about a photon's path through a double slit by using an entangled partner photon. As long as this which-way information (or the *welcherweg* information, in German) remains accessible to an observer, no interference can be detected in the patterns made by the photons going through the double slit. But if this information were to be erased, Scully and Drühl showed that one would observe interference. Their paper on the quantum eraser was published in 1982.

By 1995, Zeilinger and colleagues carried out a version of the quantum eraser experiment, as did a few other teams, but none of the experiments were quite the ideal *gedankenexperiment* that Scully and Drühl were after. Scully eventually joined hands with Yoon-Ho Kim of the University of Maryland in Baltimore, and his colleagues, and in January 2000 they published the results of an experiment that was closest in spirit to the original idea.

The experiment uses an atom that can be made to emit entangled photons when hit with a laser pulse. Imagine two such atoms, A and B. Each atom emits a pair of photons. The atoms are arranged such that one of the entangled pair of photons goes toward a screen. Let's call it the "system" photon. Both A and B can emit a system photon. The two atoms are placed side by side, such that their system photons appear to be coming through a double slit. So the double slit in this scenario is virtual; all we have are the two atoms sending out system photons. If all we had were system photons and we had no other information (so ignoring the entangled photons for now), then the system photons would create an interference pattern on the screen. That's because any system photon that lands on the screen could have come from either atom A or atom B, or from one or the other slit (assuming we have no way of telling which atom the photon came from).

But that's not all we have. For each system photon that an atom emits, it emits another photon in the opposite direction; let's call it the "environment" photon, which is entangled with its system photon. The environment photon contains information about which

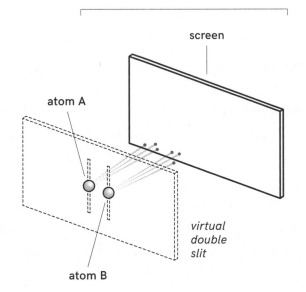

SYSTEM PHOTON

screen

atom A

virtual
double
slit

atom B

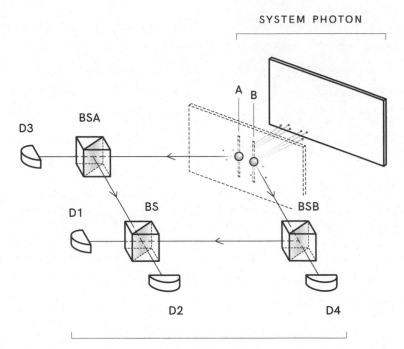

atom (or analogously, which slit) the system photon came from. The key now is to either preserve or destroy this information, and see what happens with the system photon that lands on the screen. Does it act like a wave or a particle?

Take a pair of photons emitted by atom A. The system photon goes toward the right of the screen, where it's recorded on a photographic plate. The environment photon heads left toward a set of beam splitters designed to either preserve or erase the which-way information. It first encounters beam splitter BSA. At BSA, the photon can be either transmitted to a detector D3 or reflected toward another beam splitter, BS, whereupon it can be reflected to D1 or transmitted to D2. Similarly, an environment photon from atom B will end up at D1 or D2 or D4.

It's clear that D3 will click only if the environment photon came from atom (slit) A. So a click at D3 constitutes which-way information about its partner system photon (note that this information is obtained without physically disturbing the system photon in any way). Similarly, if D4 clicks, we know the corresponding system photon came from atom (slit) B.

However, once an environment photon gets reflected at either BSA or BSB and then goes past the central beam splitter BS and hits either detector D1 or D2, it's impossible to tell whether it came from atom (slit) A or B, because environment photons from both atoms (slits) can trigger either D1 or D2. The which-way information for the corresponding system photon is erased.

Now imagine that you have collected the pattern made by a large set of system photons, and the clicks made by the environment photons. If you consider only those environment photons that ended up at D3 and D4, and looked at the pattern that the corresponding system photons made on the photographic plate, you *don't* see an interference pattern. That's because for each of those photons, we know which slit it came through. We get particle-like behavior.

But if you look at the pattern made only by those system photons whose corresponding environment photons were detected at D1, something strange happens: you see interference fringes. The same goes for environment photons detected at D2. The detection at D1 or D2 has *erased* the which-way information. At the photographic plate, there's no way to tell whether the corresponding system photons came from the left or the right slit. The paths become indistinguishable, setting the stage for superposition and for interference.

One of the most astonishing facts about the quantum eraser experiment is that the act of erasing the which-way information can be delayed for an arbitrary length of time. Say the system photons were detected almost immediately on the screen and their positions recorded. The environment photons, however, were allowed to travel, maybe for kilometers, before they went through the various beam splitters and onward to the detectors. If you analyzed the patterns made by *all* the system photons while *all* the environment photons were still in flight, you would not see any interference (because, in principle, you still have access to which-way information).

But once the environment photons encounter the beam splitters and hit the final detectors, and if you then selectively analyze the measurements already made of the system photons, you will see something entirely different. Pick only those environment photons that reach D3 or D4 and check their corresponding system photons— you will *not* see an interference pattern. But pick those system photons whose corresponding environment photons caused either D1 or D2 to click—and hence erased the which-way information—and you will see an interference pattern. Was the pattern always there? Or did it reappear?

If the idea of waiting and waiting before choosing what to do with the environment photons seems like theoretical fantasy, no one bothered to tell some physicists in Austria. A little more than a decade after Scully and Yoon-Ho Kim did their experiment, Zeilinger and colleagues were ready to test such fantasies across the mountaintops on La Palma and Tenerife.

The delayed-choice quantum eraser experiment done by Zeilinger's team is among the most sophisticated of all the variations of the double-slit experiment. Rupert Ursin, once a student of Zeilinger's, now a senior member of the team, recalled the travails of the seven-hour flight from Vienna to the Canaries. They were carrying almost two-thirds of a ton of equipment. For Europeans used to borderless travel, getting the equipment past customs in La Palma wasn't trivial. "Believe it or not, the Canaries are outside of the European Union," Ursin told me, sounding somewhat miffed. Actually, the islands are an autonomous part of Spain but still require customs checks for tax reasons.

The team had a logistics company lug the equipment up to the summit of Roque de los Muchachos, where the scientists proceeded to set up an extremely sensitive experiment. They began working cheek to jowl. "You better have good friends before you start [such an] experiment, because you'll hate them when you finish the experiment," Ursin said.

The experiment, in principle, is much the same as the one described in the previous section. But the practical details differ enormously. The experiment was spread over two physical locations: one atop the mountain in La Palma and the other near Mount Teide in Tenerife, 144 kilometers away as the crow—or in this case, the photon—flies. Most of the equipment was at La Palma, including a source of entangled photons.

In the previous experiment, two atoms were positioned such

that when one of the atoms emitted a pair of entangled photons, the system photon behaved *as if* it came through a double slit, and the environment photon went the other way, carrying information about which (virtual) slit the system photon came through. In the Canary Islands experiment, there is only one source of entangled

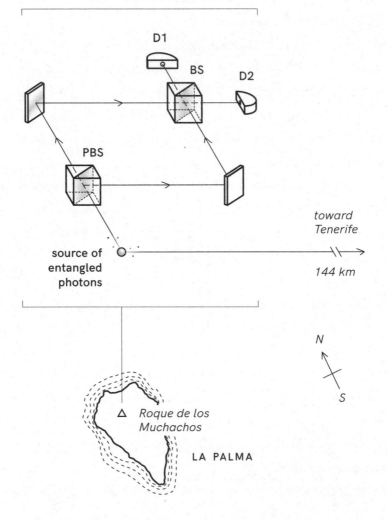

SYSTEM PHOTON

D1

BS

D2

PBS

toward Tenerife

source of entangled photons

144 km

N

S

△ *Roque de los Muchachos*

LA PALMA

photons. It emits a system photon and an environment photon. The system photon is sent into a Mach-Zehnder interferometer at La Palma and is detected immediately at either detector D1 or detector D2. The first beam splitter in the interferometer is somewhat different from the beam splitters we have seen so far; instead of randomly sending a photon one way or the other, this so-called polarizing beam splitter (PBS) sends the photon one way if it's, say, horizontally polarized, and the other way if it's polarized vertically (there are experimental subtleties about what is done to the photon after it crosses the PBS, but we can leave that aside). So, if you know the polarization of the photon, you know which path it takes through the interferometer.

The entangled environment photon, however, is sent toward a telescope at Tenerife. The photons are entangled in their polarization states. The polarization of the environment photon can be used to tell which path the system photon takes through the interferometer at La Palma. Or the polarization of the environment photon can be scrambled, which is tantamount to erasing the which-way information about the corresponding system photon. This is the quantum eraser part of the experiment.

The delayed-choice part comes in because the decision to erase or not to erase is made only when the environment photon reaches Tenerife—well *after* the partner system photon has been detected at La Palma, and thus well *after* it has ostensibly already behaved like a wave or a particle.

To precisely control the length of the two arms of the interferometer at La Palma, the team used the vibrations of a tiny piezoelectric crystal. Such delicate control is hard enough in a temperature-controlled lab in the basement of a university building. On Roque de los Muchachos, it was a phenomenal feat. The laboratory was essentially a steel shipping container that was being buffeted by winds and was subject to constant day-night temperature fluctuations. "To stabilize such an interferometer in a mountain hut [at] 2,500 meters altitude is not easy," said Ursin. "This is not a nice environment."

His colleague Xiao-Song Ma, who was also then a student of Zeilinger's, recalled how once someone merely opened the door of the shipping container and the resulting acoustic vibrations changed the interference patterns. So what did they have to do to ensure that the testbed was stable and free of noise? "Everything, literally everything," Ma told me. "Even the breath of a human being or a stamp of the feet in the lab will . . . [destroy] the interference."

The numerous beaches on the islands somewhat made up for the stress of working on the summits. The team would work through the night and go to bed at sunrise, sleep for a few hours, and then head to a beach in the afternoon. I asked Ursin about which beach they frequented. "All of them," he quipped. They did prefer, however, two in Tenerife: Las Teresitas, an artificial beach built of sand brought over from the Sahara Desert, with swaying palms and calm waters made possible by a breakwater, and contrastingly, El Bollullo, one of the island's best natural beaches.

But they had to get back to the summit before sunset and start experimenting all over again.

Transmitting the environment photon from La Palma and detecting it at Tenerife was a serious challenge. The work of aiming the source at the receiving telescope had to be done in near complete darkness. While the entanglement between the system and environment photons could survive all the optical equipment (lenses, mirrors, and the like), it couldn't survive moonlight, let alone sunlight. The photons from the moon would interact with the environment photons as they flew to Tenerife, causing them to lose their entanglement with the system photons. So the researchers worked with only stars for company in an otherwise dark sky.

To receive the photon at the Observatorio del Teide in Tenerife, the team used the European Space Agency's optical ground station, with its 1-meter telescope, ordinarily used for communicating with satellites. Near complete darkness was essential. Once, one of Ursin's colleagues stood near the source, smoking a cigarette. The infrared photons from the glowing cigarette at La Palma completely saturated the receiver on Tenerife, overwhelming the signal of the lone environment photon.

Such sensitivity could be undone by one of nature's most majestic events: Saharan sandstorms. Massive storms of fine dust blowing off the Sahara Desert can engulf the Canary Islands, obscuring even normal visibility, let alone the kind needed to do single-photon experiments in the dead of night.

But when the air is clear, then, under the cover of darkness, the telescope at Tenerife receives the photon. It's time now to either retain the which-way information or erase it. The decision to erase or not to erase is based on the output of a quantum random number generator.

If it outputs a "0," the environment photon's polarization is left untouched and the photon preserves the which-way information about its corresponding system photon at La Palma. The photon

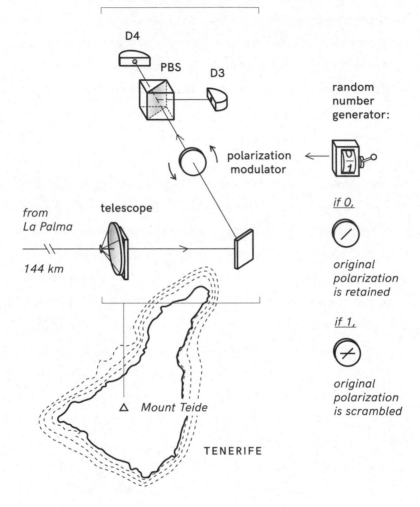

ENVIRONMENT PHOTON

D4

PBS

D3

random
number
generator:

polarization
modulator

if 0,

*original
polarization
is retained*

if 1,

*original
polarization
is scrambled*

telescope

*from
La Palma*

144 km

△ *Mount Teide*

TENERIFE

then passes through a polarizing beam splitter and ends up at D3 if horizontally polarized, and D4 if vertically polarized. Because of entanglement, we know that the corresponding system photon would have been oppositely polarized and thus we know which path it took at La Palma.

But if the random number generator outputs a "1," the environment photon's polarization is scrambled, and the which-way information encoded in the environment photon is erased. It has now a 50 percent chance of going to D3 and a 50 percent chance of going to D4. There's no way to tell whether the environment photon was horizontally or vertically polarized and so there's no way to tell which path the corresponding system photon took at La Palma.

The utterly confounding aspect of this experiment is that the measurements at Tenerife—erasing the which-way information or otherwise—are done some 0.5 milliseconds (an eternity for light) *after* the system photon has gone through the Mach-Zehnder interferometer and hit either detector D1 or D2 at La Palma. The events at Tenerife and La Palma, according to special relativity, should have no causal influence on each other. Quantum mechanics begs to differ—if one is relying on traditional notions of space and time.

We are now coming to the heart of this experiment. This intricate version of the double-slit experiment combines all the mysterious aspects of quantum mechanics: randomness, wave-particle duality, and even entanglement.

For those environment photons that were left untouched, if you look now at the clicks made at D1 or D2 by the corresponding subset of system photons at La Palma, you'll find that there was no interference; they acted like particles: half of them would have gone to D1 and half to D2 when the two arms of the interferometer were of equal length (the experimenters changed the length of one of the paths in small increments continuously, leading to different photon counts at D1 and D2, but that's a detail we can put aside).

But for those environment photons whose which-way information was erased once the environment photons were detected at Tenerife, the corresponding system photons at La Palma showed wavelike behavior: all those photons ended up at D1 and none at D2 when the path lengths were equal.

This is worth reiterating. The measurement on each system photon at La Palma is done 0.5 milliseconds *before* anything is done to the partner environment photon at Tenerife. The data on the system photon is already in the bag, so to say. Only later do we find that a subset of these photons ends up acting like particles, going through one or the other arm of the interferometer, and another subset acts like waves, with each photon ending up in a superposition of taking both paths. And because the determination of which subset does what is up to the quantum random number generator at Tenerife, if you did multiple runs of this experiment, each time a different subset of photons would show interference.

For those disturbed by the implications of the standard way of thinking about quantum mechanics, this experiment raises the stakes. First, complementarity cannot be overcome. Second, entangled or spooky action at a distance, hence nonlocality, seems to be a real phenomenon. And as tests of Bell's inequality had already showed, if quantum mechanics is complete, this seems to imply superluminal or faster-than-light signaling. Otherwise, what's being done in Tenerife to the environment photon cannot have an effect on the outcome at La Palma.

However, there is a deeper principle at stake here. Quantum mechanics is not only asking us to give up notions of locality in 3-D space but our notions of time too. The events at Tenerife, in our usual way of thinking, happen later in time, yet still influence the outcome of measurements at La Palma, even though each measurement at La Palma is done and dusted well before the partner environment photon reaches Tenerife.

Language fails us at this point. *Here and there, past and future* don't quite work.

I asked Ursin if this made him think about interpretations of quantum mechanics—the various attempts to understand what may be happening at the most basic level of reality that go beyond the Copenhagen interpretation. Ursin, however, is interested in harnessing the weirdness of quantum mechanics for technological uses. Interpretations are for old fogies. "I am the young generation, the next generation of quantum physicists," he said. "This is only a question which is interesting for gray-haired people, but I don't have so many gray hairs."

This is hardly a modern-day response. Even during the times of Niels Bohr, when Bohr was persistent in his explorations about the nature of reality, the younger physicists around him, with the exception of course of Heisenberg and Pauli, were more nonchalant. The Danish physicist Christian Møller, who was Niels Bohr's assistant at one time, said: "Although we listened to hundreds and hundreds of talks about these things, and we were interested in it, I don't think . . . that any of us were spending so much time with this thing . . . When you are young it is more interesting to attack definite problems. I mean this was so general, nearly philosophical."

Hearing Ursin talk also brought to mind something John

Wheeler wrote in one of his papers. He quoted Gertrude Stein on modern art (possibly erroneously): "It looks strange and it looks strange and it looks very strange, and then it suddenly doesn't look strange at all and you can't understand what made it look strange in the first place." For a younger generation raised on the mysteries of quantum mechanics, the strangeness may be passé.

Xiao-Song Ma, however, who is of the same generation as Ursin, has philosophical concerns. While he doesn't dismiss the Copenhagen interpretation, he hopes experiments can lead us to better ones. "I hope there will be some more intuitive interpretations of quantum physics [that are] more involved than Copenhagen," he told me. He's back in China creating ever more sophisticated experiments to further expose the apparent strangeness of the quantum world.

For him and others of his ilk, Wheeler's own words are a salve: "The final story of the relation between the quantum and the universe is unfinished business. We can well believe that we will first understand how simple the universe is when we will recognize how strange it is."

One of the key tenets of the Copenhagen interpretation is the idea of the collapse of the wavefunction, which ostensibly happens when we perform a measurement using classical instruments. These measurements are considered irreversible and they imply a boundary between the quantum and the classical. The quantum eraser experiment pushes us to reexamine our notions of what constitutes a measurement (and hence collapse) and the existence of a quantum-classical boundary.

Take the environment photon in the Canary Islands experiment. It contains information about which path the system photon took through the interferometer. Measuring the environment photon in Tenerife involved a silicon avalanche photodiode, which detects a photon by turning it into an electrical signal involving billions of electrons. The wavefunction of the system-environment photon pair is said to have collapsed at that point.

But given that there has never been an experiment that has found any physical evidence of this process of collapse, it's unclear what collapse actually means. Experimentally, what one is doing is making measurements and predicting the likely outcomes, and if those statistics are borne out over a number of identical experiments, quantum mechanics claims collapse happened in each run of the experiment. But did it?

Quantum mechanics doesn't claim collapse when the environment photon is in flight. But theoretically, one could argue for collapse, because as long as the environment photon contains which-way information that can be extracted, the system photon is going to behave like a particle. The only difference here is that the collapse can be reversed, because the environment photon is itself quantum mechanical. One can erase the which-way information, thus undoing what could have been thought of as a collapse of the wavefunction.

It's the interaction of the environment photon with the photodiode that creates a situation where the information is now entangled with billions of electrons. It's impossible to reverse the quantum states of all those electrons. This does have the whiff of an actual collapse.

Consider, however, a scenario in which the environment photon interacts with a single atom and deposits its information in the energy state of the atom. Such an atom, if held in isolation, is a quantum mechanical object, and its state can be reversed in principle and the information erased, and any earlier collapse undone. Why can't we treat the interaction with the environment photon and the atom as the boundary at which the wavefunction collapses? Well, because this particular measurement, using an atom, is reversible.

"If I replace the macroscopic detector with a microscopic [detector] that physically I do know how to reverse the evolution of, then you can show, 'Oh, look, collapse never happened,'" said Aephraim Steinberg when I met him at the University of Toronto. Steinberg is a highly regarded experimentalist who is equally at ease with the theoretical and philosophical aspects of quantum mechanics. "That's the motivation of the quantum eraser." If the information in the environment photon had actually caused a collapse of the system photon's wavefunction, then nothing could bring back the interference. But the quantum eraser allows you to do so.

The only way to test whether an actual collapse happens, thus making it impossible to reverse the state of the system, is to do an experiment where quantum mechanics claims there is a collapse—such as when a photon hits a photodiode and causes an avalanche of electrons—and then reverse the process, and somehow erase the which-way information encoded in all those electrons, and then check to see if the interference fringes come back.

If the interference doesn't come back, then one could rightfully say that the wavefunction did collapse. But such an experiment has never been done, and is likely never going to be done, because it involves the near-impossible task of reversing the evolution of macroscopic systems and then looking for interference. It'd be like trying to unscramble an egg.

So, either one has to say that collapse sometimes happens, but our technologies are unable to test if the collapse permanently destroys interference, or one has to say that the wavefunction continues to evolve according to the Schrödinger equation (the wavefunction now involves not just the system-environment photon pair but the states of the billions of electrons they engendered as well)— and that there is no real collapse.

These are the key stumbling blocks of the Copenhagen interpretation and indeed of the standard formulation of quantum mechanics. What constitutes a measurement? Where is the boundary between the classical and the quantum? What does it mean to say that the wavefunction collapses? There's an even more basic question staring at us from within the formalism: Is the wavefunction real? Does it have—as philosophers like to say—"ontological" reality?

Lev Vaidman can still recall his 1991 meeting with Avshalom Elitzur. Vaidman was working in what was for him a five-year "dead-end" program at Tel Aviv University, which involved doing research and teaching high school students. Elitzur, like Vaidman, was in his thirties. But Elitzur never finished high school and instead started teaching himself quantum physics, among other subjects (in his résumé today, there are only two entries under the heading "Education," one of which is "Autodidact"). It was while he was still a student studying philosophy of science with no high school diploma or undergraduate or graduate degree to his name that Elitzur came to

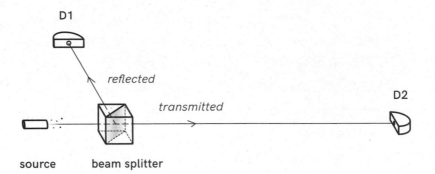

Vaidman and posed a serious question: can quantum mechanics be used to find objects without interacting with them?

As Elitzur and Vaidman figured out, the answer is yes, and the principles behind it were first identified in 1960, by German physicist Mauritius Renninger.

Consider a source of single photons aimed at a beam splitter. The photon will go toward either detector D1 or D2. But unlike the setup we saw earlier when building up to a Mach-Zehnder interferometer, the arm lengths in this setup are unequal, with D2 being much, much farther away than D1, so that it takes, say, 1 second to reach D1 but 5 seconds to reach D2. Quantum mechanics says that, until there's a measurement at either D1 or D2, the wavefunction of the photon will be in a superposition of having taken both paths. If one second later the photon is detected at D1, the wavefunction collapses; the photon is now at D1 and not at D2. Now consider the case when the photon is detected at D2. The detector will click after 5 seconds. But here's the intriguing aspect of this experiment: after 1 second, if D1 hasn't clicked, we *know* that the photon has gone the other way and is headed to D2. The negative result (the nondetection at D1 after 1 second) is already giving us information that the photon will reach D2: the wavefunction has potentially already collapsed, even though the actual measurement at D2 is yet to occur. It's the simplest example of an interaction-free measurement.

Elitzur and Vaidman applied this principle to solve what's now called the Elitzur-Vaidman bomb problem, which we encountered in the prologue. It's time to revisit it. There's a factory that's making bombs with triggers so sensitive that even a single photon hitting the trigger can cause the bomb to explode. But the factory produces some duds with no triggers. The task at hand is to separate the duds from the good bombs. You are allowed to blow up some bombs in the process. Of course, looking at the bomb to see if it has a trigger is out of the question, because looking involves shining light, and that would result in a detonation. Turns out the double slit or its special case, the Mach-Zehnder interferometer, is tailor-made for the task.

Imagine the bomb (dud or live) alongside one of the arms of the interferometer. The live bomb has a trigger, and it's the trigger that

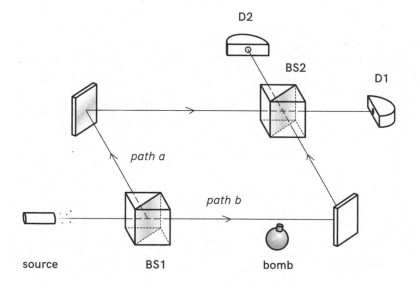

lies in a photon's path and obstructs the path. A dud has no trigger, and hence does not impede the path. For argument's sake, let's also assume that one can physically pick up and move these bombs around without exploding them: it's just photons that can blow them up (maybe they are being handled by robots in a dark room).

The duds are easy to find. The interferometer functions as if there's no obstruction in any of the paths, so the photons will be in superposition of taking both paths and there will be interference. If you send a million photons, one by one, into the interferometer (with today's technologies, that can be done in no time at all), all of them will end up at detector D1 and none will go to detector D2.

Now, if there's a live bomb in one of the paths, things change. The bomb acts like a which-way detector or a sensor for telling which path the photon takes through the interferometer. So the photons are going to act like particles: each photon is going to go

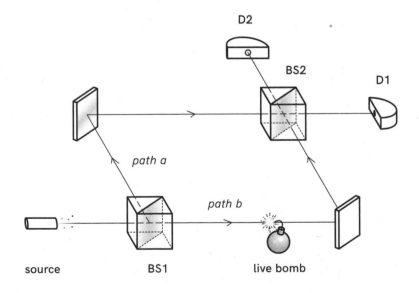

through *either* path a *or* path b (which has the bomb). There are three possible outcomes:

One is that the photon takes path b, encounters the trigger, and detonates the bomb. That's that, then—the bomb is lost, and so is the interferometer; let's assume that we can build one more in a jiffy and start all over again.

Now say the photon takes path a and encounters the second beam splitter. Because the photon is now acting as a particle, it has a 50-50 chance of exiting through either arm of the second beam splitter. So half the time the photon will go to D1. This is the second outcome. Unfortunately, D1 also clicks in the case of a dud bomb, so the result is inconclusive. We repeat with more photons for a different result.

The third outcome is the key. The photon takes path a, goes through the second beam splitter, and ends up at D2. This is a clear sign that there's a live bomb obstructing one of the interferometer paths. We know that if both paths are unobstructed, we get interference—which means all photons go to D1 and never to D2. But if D2 clicks, as in this third scenario, that's because there is no interference. The photon is taking one path *or* the other, because something is acting like a which-way measurement—in this case the live bomb. We have essentially detected the presence of the live bomb without blowing it up.

The probabilities of the outcomes with a live bomb in place are easy to compute. Half the time, a live bomb will cause the interferometer to blow up. One-quarter of the time, the photon will end up at D1—but the information is useless. One-quarter of the time, the photon ends up at D2: we know we have a live bomb in path b.

Quantum mechanics has allowed us to do something that is impossible to do with classical physics: we have distinguished a live bomb from a dud without looking at it.

Today the idea of interaction-free measurements has become relatively commonplace. In 1991, its importance was far from obvious. Elitzur and Vaidman began circulating preprints of their paper on the topic (with a section on "How to test a bomb without exploding it"), and also sent it to *Physical Review Letters*. They got back a referee report saying that while the paper was interesting, it wasn't the kind of work that *PRL* usually published. *Physics Letters A* also rejected it (the then editor of the journal told Vaidman later that the unnamed, nay-saying referee "was a very big shot").

The paper did see the light of day in 1993 in another somewhat less prestigious journal, but the idea gained prominence when Roger Penrose wrote about it in his 1994 book, *Shadows of the Mind*. Given that Elitzur and Vaidman worked in Israel, Penrose rather cheekily suggested that the experiment could be coopted for building what he called the "Shabbos switch," to help those of the Jewish faith observe Shabbat, which starts just before sunset on Friday and ends after sunset on Saturday. During this period, strict adherents are not supposed to light a fire or even turn on appliances. Penrose's Shabbos switch could help someone turn on appliances without actually doing it. Imagine replacing the bomb in the Elitzur-Vaidman experiment with your finger. Half the time, a photon entering the interferometer will hit your finger, and nothing happens. But one-quarter of the time, a photon will go through the other path, thus not interact with your finger, and reach detector D2, which could then flip a switch and turn on an appliance. "Surely . . . it can be no sin to *fail* to receive the photon that activates the switch!" wrote Penrose.

Jokes aside, interaction-free experiments highlight the disturbing conceptual questions thrown up by notions of collapse. When there's a live bomb in one arm of the interferometer, standard quantum mechanics says that the wavefunction of the photon collapses—making the photon act like a particle and go through one arm or the other. Half the time, the photon meets the bomb and blows the whole thing up. The other half of the time, it takes the bomb-free path, which allows us to infer the presence of a bomb in the other arm. But nothing interacted with the bomb. What does collapse mean in this context?

The collapse of a wavefunction has built into it the two elements that bothered Einstein: randomness and action at a distance, with the latter concerning him much, much more than the lack of determinism. In the standard formalism, when a wavefunction collapses, we can only assign probabilities to the outcomes of the collapse. The outcome is inherently random. Also, when the wavefunction involves two or more particles that have interacted at some point, meaning they are entangled, then the collapse of the wavefunction due to a measurement on one particle affects the entangled partners instantaneously—making the influence nonlocal. "Collapse has nonlocality and randomness," Vaidman told me. "This is the only phenomenon in quantum mechanics which has these two properties." Just like Einstein, Vaidman is bothered by such a theory. He'd rather see an alternative take shape. "I think a theory without action-at-a-distance and without randomness is a much better theory."

For Vaidman, interaction-free measurements are the clearest indication that any theory that invokes a measurement-induced collapse of the wavefunction cannot be the correct theory. He's not the only one to think so. Coming up with alternative interpretations or theories to explain the experimental observations has consumed the minds of a small subset of quantum theorists, a trend that began with Einstein and his insistence that there must be a theory that is both local and realistic. While Einstein's particular desire for a local realistic theory or local hidden variable theory has been refuted by the experiments done by Clauser, Aspect, Zeilinger, and others, there are alternatives very much in the running and, some would say, gaining ground, because the simple Mach-Zehnder interferometer and, by extension, the double-slit experiment continue to produce glaring paradoxes.

Lucien Hardy, whom we met in the prologue, was a PhD student in the early 1990s when he saw a preprint of the Elitzur-Vaidman bomb paper. This was before the days of the arXiv Internet server (where authors these days upload their preprints for everyone to read)—you got to read a preprint if it was sent to you. "Fortunately, they had sent a preprint to my supervisor, Euan Squires, and Euan showed me this preprint, and we both got very excited," Hardy told me.

In the Elitzur-Vaidman thought experiment, the bomb is itself a classical device. What if, thought Hardy, the bomb is quantum mechanical? What would constitute a quantum mechanical explosion? Once the thought entered his mind, it didn't take him long to devise an experiment with two Mach-Zehnder interferometers, in which an explosion happens when a negatively charged electron meets its positively charged anti-particle, a positron.

The setup is essentially two interferometers placed side by side (see next page). The first one is for electrons: a source shoots electrons one at a time into the interferometer, which have the choice of taking path a- or b- ("-" here denotes the electron's negative charge). The second one is for positrons, which can go along a+ or b+. The interferometers are arranged such that path b- of the electron and path b+ of the positron cross just before they hit their respective fully reflecting mirrors. The whole setup, in principle, is built to exacting standards, so that if an electron and a positron leave their respective sources at exactly the same time, and if the electron happens to take path b- and the positron path b+, then the two particles will encounter each other at the point where their paths intersect. This is a recipe for an explosion: a particle and its anti-particle when brought together will annihilate into pure energy.

Let's start by analyzing the electron interferometer, while ignoring its positron counterpart. We know that electrons are going to act like waves, so all of them will end up in detector C- ("C" for constructive, "-" for negative charge), while none will reach D-. Similarly, if you consider only the positron interferometer, independent of the electrons, all the positrons will reach C+, and none will reach D+.

But put them side by side as shown, and suddenly, the wave nature of the electrons and positrons sometimes disappears. That's because we now have a which-way detector built into the

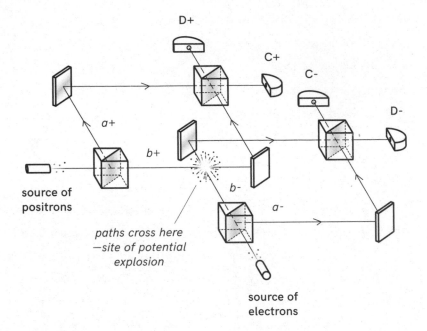

D+

C+

C-

a+

b+

D-

source of
positrons

b-

a-

*paths cross here
—site of potential
explosion*

source of
electrons

configuration of the two interferometers: it's the equivalent of having a live bomb in one arm of each interferometer.

First, let's tackle cases when the electrons and positrons behave like waves. For an electron, this happens when the positron goes through path a+. In this situation, there's no impediment to the electron's progress through its interferometer, so it'll be in a superposition of taking paths a- *and* b-, and consequently ends up at detector C-. Similarly, when an electron takes path a-, the corresponding positrons will end up at C+.

But there are times when the electrons and positrons will act like particles. Take the case of the positron going through path b+. For the electron interferometer, the positron's presence in path b+ is the same as having a detector in path b-, so the electron is going to act like a particle and take path a- *or* path b- with equal probability. If it takes b-, then the electron encounters the positron and annihilates. But if it takes a-, then at the final beam splitter, it's going to go to *either* C- or D-. When it was acting like a wave, the electron would never go to D-. So if an electron makes it to D-, it's a signal that there was a positron in path b+. An electron is able to "sense" the presence of a positron without actually encountering it—an interaction-free measurement.

Since the two interferometers are symmetric, you get the same result if you analyze the positron side of things. If an electron is in path b-, then a positron can end up at either C+ or D+; if a positron makes it to D+, it signifies there was an electron in path b-.

Hardy was not done, however. The mathematical formalism, he showed, predicts that one sixteenth of the time, on average, D+ and D- will click simultaneously. So if you did the experiment a million times, this will happen about 62,500 times. And this presents a paradox.

Here's why. From the point of view of an electron, if D- clicks, it means that a positron was in path b+. From the positron's perspective, if D+ clicks, an electron was in path b-. So, when D+ and D- go off together, it means that the electron was in b- and the positron in b+, at least according to classical logic. Which, recall, is the original recipe for an annihilation. But in these one-sixteenths of the cases, there is no annihilation: the electron makes it to D- and the positron to D+, without an attendant explosion.

If all this feels a bit like *gedanken* mumbo jumbo, and you cry out, "Surely, this can't happen in a real lab experiment," you would be wrong. A nearly exact replica of this experiment, using photons and their polarizations, was done by Dirk Bouwmeester at the University of California at Santa Barbara and his colleagues. They used photons because technology doesn't exist to do the experiment with electrons and positrons.

Hardy's paradox is real. But the paradox arises because we are talking and thinking classically of space and time, of particles taking *this* or *that* path, and reaching *this* or *that* detector. Nature has its own inimitable way of doing things.

D+ and D- go off simultaneously sometimes because, according to the formalism of quantum mechanics, the particles are somewhat entangled just before they hit their respective final beam splitters. What's even more astonishing is that the entanglement is between an electron and a positron that came from different sources. In the previous experiments, when two photons were entangled, they had been emitted by the same atom, or had in some way interacted and gotten entangled. Not so in this case.

So if we hold on to local realism—the bedrock of classical physics—and give a local description in terms of particles taking actual paths, then we end up with Hardy's paradox. But let go of local realism and we are faced with an unavoidable conclusion: "Quantum theory is nonlocal," said Hardy. His paper, as it happened, was published in 1992 in physics' premier journal, *Physical Review Letters* (which had rejected Elitzur and Vaidman in 1991); the irony is that Hardy's work was inspired by the Elitzur and Vaidman paper that was yet to see the light of day in a peer-reviewed journal. In his paper, Hardy acknowledged the Israeli duo's paper as a "Tel Aviv Report, 1991." In 1994, physicist David Mermin wrote that Hardy's thought experiment to demonstrate nonlocality is "simpler and more compelling" than tests of Bell's theorem: "[It] stands in its pristine simplicity as one of the strangest and most beautiful gems yet to be found in the extraordinary soil of quantum mechanics."

The experiments in this chapter illustrate the counterintuitive nature of the quantum world, even more so than the wave-particle duality we encountered earlier. But what will become clear in the forthcoming chapters is that phrases and words like *wave-particle duality, nonlocality, spooky action at a distance, superposition of being here and there, the randomness of nature, nondeterminism*—these are ways of thinking about what's happening in the quantum realm when the mathematical formalism is interpreted according to the Copenhagen interpretation. Other interpretations, sometimes with different formalisms, sometimes just reinterpretations of the same formalism, give us a very different view of the quantum underworld.

At the heart of the formalisms is the wavefunction. What do we make of it? Does it merely represent our knowledge about the quantum world, making it epistemic? Or is it something real (as potentially evidenced by interaction-free measurements, which suggest that it's the wavefunction that's "sensing" the bomb, for example), making the wavefunction a key ingredient of reality and part of the ontology of the world? And regardless of whether it's ontological or epistemic, what does one make of the wavefunction's collapse?

As successful as quantum mechanics is as a theory—and it's by far the most successful physical theory we have—such questions continue to haunt those who ponder the very essence of reality. For them, some of whom *are* gray-haired, it's not enough to harness quantum mechanics to build better technology, to "shut up and calculate," as the saying goes. Among the earliest thinkers looking for a deeper description of reality, of course, were Einstein and de Broglie. After them, the first physicist to seriously tackle the conceptual difficulties of the Copenhagen interpretation by formulating a hidden variable theory was an American physicist who briefly worked alongside Einstein at Princeton before he was hounded out of an America steeped in McCarthy-era paranoia.

6

BOHMIAN RHAPSODY

Obvious Ontology Evolving the Obvious Way

There's an entirely different way of understanding all this
stuff (a way of being absolutely deviant about it, a way of being
polymorphously heretical against the standard way of thinking, a way of tearing quantum mechanics all the way down
and replacing it with something else).

—**David Albert**

It's no small irony that one of the strongest moves against the
Copenhagen interpretation was made by a student of Robert
Oppenheimer. Best known to the world as the scientific director of
the Manhattan Project, the US effort to build the atomic bomb,
Oppenheimer was a strong proponent of Niels Bohr's view of the
quantum world. He founded the first school of theoretical physics
in the United States and taught quantum mechanics at the University of California, Berkeley, where "Bohr was God and Oppie was
his Prophet." A young David Bohm came to do his PhD with

Oppenheimer and was probably deeply influenced by Oppenheimer's evangelism of Bohr's ideas. But there was already a hint of a rebel in Bohm's behavior.

World War II was devastating countries, and America was building the bomb. Bohm became a member of the Communist Party and got involved with union activities—which meant that he could not get the security clearance necessary to defend his thesis, which was on a topic considered sensitive enough to be classified. Eventually, Bohm got his PhD, but only after Oppenheimer reassured UC Berkeley that his student's thesis deserved a degree without the customary defense.

Soon after he got his PhD, Princeton University gave Bohm a job (after all, he was one of the brightest of the crop of young American theorists, and "probably Oppenheimer's best student at Berkeley"). Bohm began teaching quantum physics. But soon his past caught up with him. In 1949, the House Un-American Activities Committee subpoenaed him to appear before Congress and talk about his and his colleagues' communist connections. Bohm refused, and pleaded the Fifth. This was contempt of Congress: Bohm was indicted, arrested, but then released on bail. A court subsequently acquitted him, but the damage was done. Princeton suspended him and barred his access to university facilities, and when his contract came up for renewal in 1951, they demurred.

But Bohm hadn't been sitting on his hands. In 1951, he published one of the most lucid textbooks on quantum mechanics, *Quantum Theory*, in which he elegantly explained the Copenhagen viewpoint (the book was a result of his pedagogical efforts at Princeton). It was also the book in which Bohm reformulated the

Einstein-Podolsky-Rosen (EPR) thought experiment, crystallizing its essence better than Einstein himself. After the book came out, Bohm and Einstein met and discussed quantum mechanics, a discussion that played a key role in Bohm's evolving views about the nature of reality.

But before he could do that, his career took a turn for the worse. When his contract at Princeton wasn't renewed, Bohm knew his days as an academic in the United States were numbered. He moved to Brazil in October 1951, where a coterie of former Princeton graduates got him an academic appointment at the University of São Paolo, with recommendations from no less than Einstein and Oppenheimer. Bohm was looking forward to collaborating with physicists in Europe, but those hopes were dashed when the US State Department confiscated his passport. Bohm was now officially in exile in Brazil and he would stay there until 1955, when he would leave for Israel.

In the meantime, Bohm published a paper that challenged the anti-realist stance of the Copenhagen crowd. It seemed to come out of the blue, but in hindsight, his 1951 textbook contained hints of his radical ruminations. In the book, he openly discussed the idea of hidden variables. Using the laws of thermodynamics to make his point, he argued that the reason why we have to deal with probabilities of outcomes in thermodynamics is because we don't have complete knowledge of the properties of, say, the underlying molecules of some gas. Variables that capture these properties would constitute hidden variables. Could the probabilities that arise in quantum theory—for example the probability of finding an electron here or there—be similarly the outcome of not knowing enough about variables that capture the properties of some hidden layer of reality?

Even though he raised these issues in the book, Bohm wasn't yet convinced that the Copenhagen interpretation needed rethinking. "Until we find some real evidence for a breakdown [of quantum theory] . . . it seems, therefore, almost certainly of no use to search for hidden variables. Instead, the laws of probability should be regarded as fundamentally rooted in the very structure of matter," he wrote. So Bohm, while pondering heresies, was still espousing Bohr's views in his book (the way Zeilinger still does). In fact, Bohm's book was not "only orthodox in the Copenhagen sense but one of the clearest and fullest, most penetrating and critical presentations of the Copenhagen view ever published." He even went so far as to say that the "general conceptual framework of the quantum theory cannot be made consistent with the assumption of hidden variables." He used the EPR result to make his case. As Einstein, Podolsky, and Rosen had pointed out, their thought experiment suggested that under assumptions of locality, both the momentum and position of two entangled particles would have clearly defined values—but this would contradict the uncertainty principle, which Bohm called "one of the most fundamental deductions of the quantum theory."

Therefore, Bohm concluded, "no theory of . . . hidden variables can lead to *all* of the results of the quantum theory."

But all that changed a year later. In 1952, Bohm published his seminal paper in *Physical Review*, titled "A Suggested Interpretation of the Quantum Theory in Terms of 'Hidden' Variables." (The paper acknowledges only one person: "The author wishes to thank

Dr. Einstein for several interesting and stimulating discussions.") It was the first and clearest example of a theory that Bohm himself had said could not be conceived. It also implicitly showed that John von Neumann had been wrong: it was possible to come up with a theory with hidden variables that could explain experimental observations in quantum physics and recover realism and determinism.

It's in the nature of the debate about quantum theory that the proponents of the Copenhagen view are not particularly vocal. They don't have to be. They have history on their side. Niels Bohr, Werner Heisenberg, Wolfgang Pauli, and many other giants of theoretical physics have already argued the case for the Copenhagen interpretation. But for some theorists thinking about the foundations of quantum mechanics, it's far from a done deal. They have to, however, raise their voices to make themselves heard, and they are usually far more passionate than the adherents to the orthodoxy. Sheldon Goldstein is no exception.

As with Bohm, Goldstein too began his career advocating Bohr's views. He was studying at Yeshiva University in New York in the late 1960s and early '70s. "I was a fairly strong defender of the Copenhagen interpretation, to the extent that I understood it," Goldstein told me when we met at Rutgers University in New Jersey on a miserably rainy day. He invited me into his long, narrow office—one side of which was lined with bookshelves filled with books on quantum mechanics. Through the large windows at the far end of the office, I could see the gray sky and the occasional skein of geese flying past. Stuck to the bookshelves were clippings of newspaper articles about Alan Sokal, a New York University professor who in 1996 duped a social studies journal into publishing what turned out to be gibber-ish, to prove a point that such journals would publish nonsense. A white T-shirt hung from one of the bookshelves; it had an image of Bohm, with the words *David Bohm, Keepin' it real*. Goldstein sat down on his swivel chair, leaned back, and put his legs up on the table, clasped his hands behind his head, and proceeded to talk for two hours, getting up only to scribble equations on the blackboard or to grab a much-thumbed copy of *Speakable and Unspeakable in Quantum Mechanics* by John Bell, its dust jacket in tatters, and quote entire paragraphs to me.

"I wanted Bohr and Heisenberg and orthodox quantum theory to be right, and Einstein to be wrong," he said.

"You wanted that?" I asked.

"Yeah, I wanted Einstein to be wrong," said Goldstein. "I'm not too proud of that, by the way. I was excited about the quantum rev-olution, and Einstein was presented as somebody who wanted to go back to old-fashioned classical ways of thought. He just couldn't get with the new modes of thinking; he was too old."

Goldstein then expressed some remorse for those thoughts about Einstein. "I think that was very unfair, but anyway, that's what I thought then," he said. "You could say I wasn't smart enough to see what a bunch of crap that was, so I swallowed it. I thought if I learned the mathematics better and looked into it carefully, I would really understand it all one day. [But] the more I learned, the more clear it became that we were all hoodwinked."

Strong words, but not unusual from those who have developed a distaste for the orthodoxy.

As Goldstein probed further into the mathematics of standard quantum theory, he was unable to make sense of what it's about. What are the fundamental entities of reality? Is it a theory about particles? Is it about waves? Is it a theory of measurements and observations? Is it a theory of wavefunctions? Is the wavefunction ontic (meaning it is *something*) or is it epistemic (in that the wavefunction represents our knowledge about something); is the wavefunction objective or subjective?

Goldstein wasn't done expressing his concerns about orthodox quantum mechanics. "Are there particles before you look? Do they have positions before you look? According to textbook quantum mechanics, presumably not. Then what do you have before you look? Or does looking create reality? Is that clear from the usual theory, textbook theory? No, it's not."

Goldstein used the double-slit experiment to further make his point about the "reality" of the wavefunction. "I don't see how you can understand the interference unless you take seriously that you have a wavefunction, an objective thing in the world, which has these two pieces, one going through the upper slit, and one going through the lower slit, and they interfere with each other," he said.

Disenchanted with the Copenhagen interpretation, Goldstein turned to work by a mathematical physicist at Princeton named Edward Nelson, who had proposed a theory called stochastic mechanics to arrive at a realistic theory of the quantum realm. The theory had actual particles in it, with positions and momenta, and these particles were being randomly buffeted by the wavefunction—resulting in a sort of Brownian motion. It wasn't deterministic and it reproduced the results of standard quantum theory, albeit after many mathematical contortions. Goldstein found it enticing but soon realized that it was too complicated and that there was something simpler hiding in Nelson's proposal.

And even as he started figuring out the simpler idea, he had a vague notion that "there was this guy David Bohm" who had proposed a deterministic formulation of quantum theory, one with hidden variables. Goldstein discovered that the idea that he was playing with—making Nelson's stochastic mechanics simpler and deterministic—was exactly what Bohm had already clearly elucidated. Here was an alternative to the Copenhagen view of things: a deterministic theory of particles that move around because of interactions with the wavefunction, which in turn is a "real" thing and evolves according to the rules of the Schrödinger equation.

Bohm's theory has a definite ontology: the world is made of particles and wavefunctions, even if wavefunctions are not "physical" in the sense that particles are physical, but nonetheless are real, objective aspects of nature. A particle has a definite position at all times, which means it has a trajectory—in direct contravention of the Copenhagen view of reality. The particle is "guided" by the wavefunction, and thus influenced not just by the usual forces (such as electromagnetism), but by a "quantum potential," a new force felt by the particle because of its interactions with its wavefunction. Moreover, the theory is deterministic: given a particle's position and its wavefunction, you can predict the particle's position at some later time. And even more emphatically, the particle's trajectory is objective reality—it exists independent of an observer.

And what of hidden variables? In Bohm's theory, the much-maligned hidden variables are nothing other than the positions of particles. To those who think Bohm is right, it's an irony that this rather obvious property has to be called "hidden": it's called so because it doesn't appear in the standard formalism of quantum mechanics, unless "observed."

Just as Goldstein discovered that his nascent ideas had already been worked out by Bohm, Bohm would discover that his theory wasn't entirely novel either. Louis de Broglie, the young French prince whom we encountered earlier, had made the first clear attempt at a theory that incorporated both realism and determinism, back in the 1920s. Recall that in 1924, de Broglie came up with the theory that particles of matter such as electrons had wavelike properties. Then in 1927, de Broglie presented another radical idea at the Fifth Solvay Conference in Brussels—that reality is made of particles and that these particles are being guided by a "pilot wave," which behaves like a wavefunction and evolves according to a form of the Schrödinger equation. So de Broglie was proposing that reality isn't wave *or* particle, as Bohr was arguing, but rather it's wave *and* particle. At the Solvay meeting, Wolfgang Pauli—who sided with Bohr—ripped into de Broglie's theory, claiming to point out certain experimental situations that it couldn't explain. A disheartened de Broglie gave up on the pilot-wave theory, and actually became a supporter of the Copenhagen interpretation.

Until, that is, Bohm entered the picture. Bohm, unaware of de Broglie's work, had reinvented the theory, but with far greater conceptual and mathematical clarity. Einstein and Pauli both alerted him to de Broglie's work. Pauli, in particular, raised some of the same issues that he had brought up after de Broglie's presentation in Brussels. But Bohm, unlike de Broglie, did not back down. He revised his draft to address Pauli's concerns and sent it to Pauli, who apparently did not read it because it was too long. Bohm wasn't amused. He sent Pauli a rather stern note: "If I write a paper so 'short' that you will read it, then I cannot answer all of your objections. If I answer all of your objections, then the paper will be too 'long' for you to read. I really think that it is your duty to read these papers carefully."

As for giving de Broglie his due, Bohm did so somewhat reluctantly. In his 1952 paper, he acknowledged that he had been alerted to de Broglie's work *after* he had completed his paper, and that de Broglie had abandoned his approach following criticism from Pauli and after de Broglie had himself realized what he took to be some of the theory's shortcomings. "All of the objections of de Broglie and Pauli could have been met if only de Broglie had carried his ideas to their logical conclusion," wrote Bohm.

He put this argument rather more colorfully in a letter to Pauli: "If one man finds a diamond and then throws it away because he falsely concludes that it is a valueless stone, and if this stone is later found by another man who recognizes its true value, would you not say that the stone belongs to the second man? I think the same applies to this interpretation of the quantum theory."

To Bohm's credit, he did push the ideas to their logical conclusion and the result was the first deterministic, realistic, hidden variable quantum theory. As Bell subsequently said, Bohm had done the impossible.

Today, the pilot-wave theory is often referred to as the de Broglie-Bohm theory. De Broglie, once he became acquainted with Bohm's work, left the Copenhagen camp and started working on a variant of his own idea called the double-wave solution, something he had started on in 1926 but had given up as being too difficult.

After decades in exile, both the de Broglie-Bohm pilot-wave theory (which Goldstein favors) and de Broglie's double-wave solution are getting some attention and even support. The latter from an unlikely group of researchers studying how droplets of silicone oil bounce on a vibrating surface of the same oil. What, you might ask, has that got to do with quantum physics?

As a graduate student, John Bush felt the same annoyance with quantum mechanics as he did with Sunday school growing up in London, Ontario, Canada. There were some questions that were off-limits when it came to religion. Unfortunately for him, he encountered similar attitudes when learning quantum mechanics and the Copenhagen interpretation. "You are telling me that the particle doesn't exist unless you observe it?" he'd ask. The instructor would go, "You can't ask that question." Bush felt he was back in Sunday school.

It was galling to Bush that human observers could somehow be held responsible for creating quantum reality. It still galls him. "This is the latest in the long line of epic human intellectual follies that have resulted from man putting himself at the center of the universe," he said when we met at his office at MIT. "It strikes me as nonsense."

Disillusioned with quantum mechanics, Bush ended up studying fluid mechanics. Little did he know that his chosen field would lead him back to quantum mechanics. The impetus came from the

2006 work of two French researchers, Yves Couder and Emmanuel Fort. They had conjured a curious setup. Imagine a petri dish filled with silicone oil that's being vibrated up and down. These vertical vibrations of the bath of oil are kept below what's called the Faraday threshold for the fluid. Above this threshold, waves form on the surface, but below the threshold, the surface remains smooth, even though there is vibrational energy in the fluid. The researchers discovered that if they let a millimeter-size droplet of the same oil fall onto the vibrating surface, the droplet would keep bouncing and begin wandering across the surface.

Here's why. A thin cushion of air between the droplet and the surface prevents the droplet from coalescing into the oil bath. Upon first impact, the vibrating surface gives the droplet a vertical kick, causing it to bounce up. The impact also creates a small wave on the bath surface. When the droplet falls back onto the surface, it encounters this wave. This time, the droplet gets both a horizontal and a vertical kick, and the process now keeps repeating. The droplet starts "walking" over the surface, guided by the very wave it creates and sustains with each bounce. The wave dictates the droplet's speed and direction.

The analogy with the theories of de Broglie and Bohm is hard to ignore. The droplet is a particle being guided by its pilot wave. What else can one do at this stage but carry out a version of the double-slit experiment?

Couder and Fort did just that. They made a barrier with two openings and submerged it fractions of a millimeter below the oil surface, such that anything moving on the surface would be influenced by the barrier. The subsurface barrier made for a double slit.

When the walking droplet approached the barrier, it went over one or the other opening (like a particle going through one slit or the other). The attendant pilot wave, however, spanned both openings and thus went over both. When the wave emerged on the other side of the submerged barrier, it was now the outcome of the interaction between two diffracted waves, each influenced by one subsurface slit. This more complex wave now guided the bouncing droplet away from the barrier. For each run of the experiment, the droplet went to a different location on the far side. The researchers collected seventy-five such trajectories, and their initial analysis showed that the droplets were going to some places and not to others—suggestive of an interference pattern. Despite there only ever being one particle-like droplet in the apparatus at any one time, its accompanying pilot wave was causing the droplet to behave as a wave. If you didn't know about the pilot wave, you'd think the droplet had gone through both slits and was interfering with itself.

Had Couder and Fort done an actual double-slit experiment using a bouncing droplet of silicone oil? Had they found a classical analogue of what happens in the quantum world? Other teams raced to duplicate the results, but failed. One team was led by Tomas Bohr, Niels Bohr's grandson, at the Technical University of Denmark near Copenhagen. Another was led by John Bush and his team at MIT. Their results revealed inadequacies in the experiment done by Couder and Fort. Bohr and colleagues showed that the French team's statistics were inadequate—seventy-five trajectories were just too few to make strong claims about what the droplets were doing. And Bush's team pointed out that the French experiment hadn't been adequately sealed off from environmental influences, so the droplet's patterns on the surface, for example, may have been affected by ambient air currents.

When the MIT researchers did a more rigorous version of the experiment, they did not see the double-slit interference pattern. Nor did they see the kind of diffraction patterns expected when a particle goes through a single slit. They attribute this to "boundary conditions": the interaction of the droplets and the waves with the walls of the petri dish, for example, making it difficult to reproduce the conditions that would be experienced by, say, a photon going through a double slit, where there are no such boundary effects. Maybe future experimentalists can come up with walking-droplet setups that negate any effects of physical boundaries. "Our results do not close the door on the quest for diffraction and interference of walking droplets," Bush's team concluded in one of its papers.

But Bush told me that they do see the kind of mystery highlighted by Richard Feynman's analysis of the double-slit experiment. In the quantum mechanical version, when both slits are open, the particle goes to certain places on the far side and not to others. Close one of the slits, and the behavior of the particle changes, as if the particle senses the closing of one slit. The classical walking droplet does the same, even though it does not exactly replicate an interference pattern in where it goes and does not go. It's fair to say that the droplet, which is going over only one opening in the barrier or the other, nonetheless can "sense" whether both slits are open or not. "Our system has that feature, if that's the mystery," said Bush.

Bush thinks the walking-droplet setup is an important classical analogue of a quantum mechanical system. They may not have replicated the double-slit experiment yet, but they are seeing phenomena that are too suggestive to ignore. For example, when they follow the seemingly chaotic movements of a droplet in the circular bath, over time its statistics resemble those of an electron moving inside a quantum mechanical corral of atoms.

Bush explains this result using de Broglie's double-wave solution. De Broglie revisited this idea, which he had abandoned after the 1927 Solvay Conference, when Bohm revitalized the single pilot-wave theory in 1952. The double-wave solution is what it says: there are two waves involved in guiding a particle. One is a localized wave, and the particle is centered on that wave and is guided by it. This particle-wave combination gives rise to another wave that behaves like the wavefunction in orthodox quantum mechanics.

According to Bush, the vibrating oil bath and the walking droplet physically replicate this two-wave system. The bouncing droplet creates and sustains the pilot wave. This wave is localized and the droplet is centered on the wave. The interaction of the droplet and pilot wave with the geometry of the vibrating surface also creates another wave pattern whose properties emerge over time and mimic those of a wavefunction. "Now we have a macroscopic realization of the physical picture suggested by de Broglie, and it exhibits many of the allegedly inscrutable features of quantum mechanics," Bush told me. "That's a hell of a coincidence."

There's a chance that that's all it might be: a coincidence. Bush isn't overly concerned. The main thrust of his argument is that physicists have to challenge the Copenhagen interpretation, and anything that gets them to do that is worth the effort. "That's why I'm a believer in this venture, even if its sole result is to get young people to question their views on quantum mechanics," Bush said.

To other quantum physicists, even those who are anti-Copenhagen, the idea that a classical system can replicate all the features of quantum mechanics is a hard sell. Goldstein, for one, thinks that the walking droplets can never replicate the key feature that distinguishes the quantum world from the classical: nonlocality, which depends on the wavefunction that, for a system of two or more particles, doesn't live in the familiar three-dimensional space of physical things.

Goldstein has an acronym for the Bohmian view of the quantum world (a "terrible acronym," he acknowledged): OOEOW. It stands for the "obvious ontology evolving the obvious way." Since Bohm came up with his formulation of quantum theory in 1952, there have been many tweaks to the theory, most notably by Bohm himself in collaboration with Basil Hiley, who worked with Bohm during the final phase of Bohm's career at Birkbeck College in London (Bohm moved from Israel to the UK and stayed put). Others who have contributed to the growing understanding of Bohm's ideas include English physicist Peter Holland and the team of Goldstein, Detlef Dürr, and Nino Zanghi. While the details differ (and contentiously so), the essence is captured by Goldstein's acronym. Dürr coined the term *Bohmian mechanics*. It's a term that Hiley does not like, but we'll stick to it in this chapter, to avoid getting tangled up in the subtleties of the different points of view. So, according to Bohmian mechanics, the quantum world consists of particles with definite positions and a wavefunction that guides these particles. If you have a system of N particles, each particle has a position. There is, however, only one wavefunction, and each particle is being influenced by it. The particles have coordinates in three-dimensional space, but the wavefunction doesn't operate in the same 3-D space. Instead, it does so in something physicists call configuration space. Take just two particles. Each particle has a coordinate in 3-D space that describes its position (x1,y1,z1 for particle 1 and x2,y2,z2 for particle 2). The wavefunction for the two particles, however, needs all six numbers (x1,x2, y1,y2, z1,z2) to describe the state of the system, and thus has to contend with six mathematical dimensions. The actual positions of the two particles in 3-D space correspond to one point in the 6-D configuration space.

It's easy to see how the number of dimensions of the configuration space quickly mushrooms as the number of particles increases—making it impossible to visualize. Nonetheless, no matter how many particles there are in the system, their individual positions in 3-D space ultimately correspond to just one point in the 3N-dimensional configuration space, where N is the number of particles.

Despite this mathematical abstraction, the wavefunction is real in Bohmian mechanics. It propagates in configuration space and evolves according to the Schrödinger equation. And it simultaneously exerts an influence on each and every particle. The wavefunction determines the trajectory of every particle. And because this interaction between the wavefunction and the particles is not taking place in 3-D space but within the confines of configuration space, the interaction is instantaneous. This is how nonlocality is built into the very fabric of Bohmian mechanics. Many have argued that this profound nonlocality (a particle in some distant galaxy, in principle, is instantly influencing a particle here on Earth) makes Bohm's ideas untenable. Even though Bohm was aware of the nonlocality built into his theory, it was John Bell who was the first to give it serious thought. He wondered if he could get rid of it. "He proved you couldn't," Goldstein said.

The tests of Bell's inequality carried out by Clauser, Aspect, Zeilinger, and others have ruled out local hidden variable theories. The correlations in the outcomes of measurements carried out on, say, two entangled particles cannot be explained by theories that posit local hidden variables, meaning variables that are not present in standard quantum theory and whose values evolve in a local manner, unaffected by what's happening at a distance. These tests, however, do not rule out *nonlocal hidden variable* theories, of which Bohmian mechanics is an example. The position of a particle, the hidden variable in the theory, is nonlocally influenced by the positions of all other particles, mediated by the wavefunction. Bell was keenly aware that his theorem did not address nonlocal theories such as Bohm's. "In fact, Bell repeatedly stressed that any serious version of quantum mechanics, and even orthodox quantum mechanics, must be nonlocal," Goldstein told me.

So, even though orthodox quantum mechanics wins out when it's pitted against local hidden variable theories, the outcome of a tussle between the orthodoxy and Bohmian mechanics is far from resolved. It has proven impossible to experimentally disprove Bohm's ideas, because the theory makes exactly the same predictions as orthodox quantum theory.

Bohmian mechanics makes a case for itself in other ways. Some aspects of quantum theory, which are axiomatic in standard quantum mechanics, can be derived in Bohmian mechanics. For example, the uncertainty principle can be shown to be an outcome of not knowing enough about the exact initial conditions of a system. So, despite the fact that Bohmian mechanics is deterministic, inadequate knowledge can make precise predictions impossible. This is not unlike chaos theory in classical mechanics: a small initial perturbation can lead to wildly different eventual outcomes in the evolution of a chaotic system (such as weather), making the system appear nondeterministic. In fact, even with all the challenges to accurate weather prediction, it is nonetheless a deterministic system.

Even the collapse of the wavefunction, which is such a bugbear in standard quantum theory—in that we don't know what it really means or whether a physical collapse actually happens—can be deduced in Bohmian mechanics. Consider Schrödinger's cat. It can be described by the positions of the "N" particles that make up the cat and its wavefunction. In Schrödinger's thought experiment, when the cat ends up in a superposition of being dead and alive, orthodox quantum mechanics requires a measurement (or observation) to collapse the wavefunction from its state of superposition to one in which the cat is either dead or alive. In Bohmian mechanics, the overall system ends up in either a cat-dead *or* cat-alive state regardless of measurement. The N particles that make up the cat will be in one configuration or the other. The observer merely discovers the state. There is no collapse of the wavefunction. The part of the wavefunction that captures the cat-dead state and the part of the wavefunction that captures the cat-alive state diverge in configuration space and no longer influence each other. There is an effective collapse—but nothing that is mysterious. The physical process is clearly mapped out by what happens to the particles that make up the cat. In addition, if one focuses on what should be regarded as the wavefunction of the cat after the experiment, one finds an actual collapse and not merely an effective one.

"So Bohmian mechanics is not a repudiation of the rules of quantum mechanics," said Goldstein. "It's simply a clarification of them. You understand where they come from, you understand more clearly what they say."

Despite such claims, support for Bohm's ideas was hard to come by. Those who favored the orthodox interpretation even pointed out that the very strengths of Bohmian mechanics—its clear ontology and the fact that particles have trajectories—were leading to problems. These problems were laid bare using—what else?—the double-slit experiment. Theorists claimed that the trajectories predicted by Bohmian mechanics to describe particles passing through the double-slit apparatus made no sense. They dubbed them surreal, a term they used to deride Bohmian mechanics.

For experimentalists, proving the existence of such surreal trajectories was going to be a challenge. First they had to find a way to determine trajectories in general, before they could focus on surreal ones. It required a whole new way of thinking about measuring trajectories. After all, traditional quantum mechanics eschewed the very notion of trajectories.

Bohmian mechanics, however, says that particles have well-defined trajectories. Chris Dewdney can recall the first time he saw the trajectories of particles going through a double slit. It was the late 1970s. He had just applied to do a PhD with David Bohm at Birkbeck College. Basil Hiley replied instead and said he'd take on Dewdney as his student. "Okay, close enough," Dewdney recalled thinking.

While he was looking around for a thesis topic, Dewdney stumbled on a book at a local bookstore on quantum mechanics by Frederik Belinfante, which had a chapter on Bohm's hidden variable theory. Belinfante suggested the possibility of calculating the Bohmian trajectories of particles in a two-slit experiment. Dewdney recalled being puzzled. "I thought, 'This is very strange.' At Birkbeck, nobody was talking about this," he said, despite the fact that Bohm himself was at Birkbeck. Dewdney talked it over with Hiley and Chris Philippidis in a coffee lounge at Birkbeck, and they decided to plot the trajectories. Today, our smartphones would make quick work of such calculations. But four decades ago, they needed a supercomputer, especially for the graphics. The programming was done using punch cards. You had to submit the cards and wait. The results came back in a small film canister. "You had to get them printed or hold them up to the light," said Dewdney, who is now at the University of Portsmouth in the UK. And when they did, the particle trajectories were clearly visible. "It was amazing, totally amazing," Dewdney told me. A particle went through one slit *or* the other, and then zigzagged its way to the screen on the far side. Taken together, trajectories bunched up in ways that eventually mimicked an interference pattern.

The trio published a paper showing the trajectories in 1979, but measuring them remained a pipe dream. "Mostly everyone in the community had believed that these are trajectories that you cannot directly measure," experimentalist Aephraim Steinberg of the University of Toronto told me.

That's because a traditional "strong" measurement, which ostensibly collapses the wavefunction of the thing being measured,

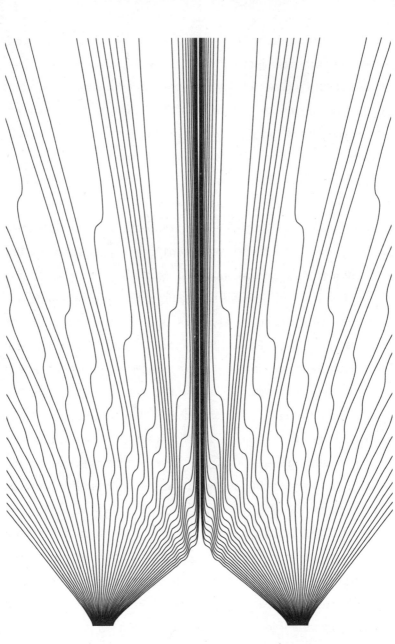

slit 1 slit 2

destroys the coherent superposition of the particle. The particle is irrevocably disturbed, even destroyed, by a strong measurement.

So, finding a particle's possible trajectory using such measurements is an impossibility. Think of how different this is from determining the path of, say, cars on a highway. If you put cameras every 100 meters to record the passing of cars, the information from these cameras can be used to reconstruct the trajectories of the cars moving past. But if you try to do this for photons or electrons, it doesn't work. Each strong measurement that tries to find out the location of a particle gives the particle such a kick that it no longer goes where it would have gone had there been no measurement. There's really no way to measure trajectories of particles using strong measurements without altering them.

In 1988, Yakir Aharonov (one of Bohm's students), along with David Albert and Lev Vaidman, came up with a theory of what they called "weak measurements." What if one doesn't try to find out the precise value of some property of a quantum system but, rather, probes it ever so gently, so as to not disturb the particle, allowing it to continue on its trajectory as if nothing happened? It turns out that the outcome of any such individual measurement is rather useless. The uncertainty in the measurement means the result could be widely off the mark. But Aharonov and colleagues showed that if you do such measurements on a large ensemble of identically prepared particles, then although each measurement alone isn't revelatory, taken together they are. The team argued that the average value of all measurements—say, in this case, of a particle's position—is an indication of its average position.

There's plenty of controversy about whether this average value is indeed giving you any relevant information about a particle's property. But some physicists saw in weak measurements a way to measure particle trajectories. In 2007, Howard Wiseman of Griffith University in Brisbane, Australia, showed that you could use weak measurements to seemingly measure the positions and momenta of particles moving through a double-slit apparatus. The idea is simple: you take hundreds of thousands or even millions of identically prepared particles and send them through the double slit one by one, and you perform weak measurements, say, at different locations between the double slit and the screen where the interference is observed. These weak measurements can then be used to reconstruct the trajectories of particles traversing the apparatus. "It must be emphasized," wrote Wiseman, "that the technique of measuring weak values does not allow an experimenter . . . to follow the path of an individual particle." That would be a violation of the rules of quantum mechanics. Nonetheless, one could, in principle, reconstruct average trajectories.

Until Wiseman's paper, the idea of measuring trajectories had been anathema, but his work changed minds. Aephraim Steinberg's team took on the task. As always with such experiments, it involved some very sophisticated optics. Even so, conceptually, the experiment is easy to understand. Steinberg's team sends each photon into a beam splitter, which steers the photon into one optical fiber or another, with equal probability. The fibers are set up to hit a mirrored prism that reflects the photons at right angles (with the left fiber hitting the left prism and the right fiber hitting the right prism). The net effect of this arrangement is the two prisms act as two slits.

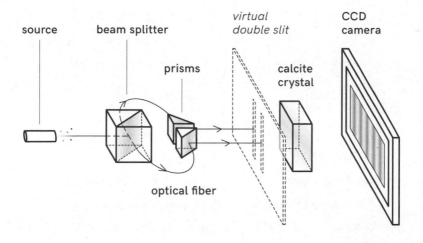

A CCD camera placed on the far side of the prisms records the photons. For each photon that lands on the camera, there is no way to tell which prism (or slit) it came from. This indistinguishability leads to interference, which is captured by the camera.

The innovation was in a block of calcite crystal placed between the double slit and the CCD camera. The photons have to traverse the calcite crystal, which has the property of rotating the angle of polarization of the photon moving through it. By carefully aligning the crystal, it's possible to use this change in the photon's angle of polarization to get a sense of the direction in which the photon is moving, relative to the midline. This is a weak measurement: the change in polarization is tantamount to catching a whiff of the propagation direction without destroying the photon. It's a measure of the angle at which the photon is traveling and hence a proxy for its momentum. Of course, then to actually measure the change in polarization requires a strong measurement, which destroys the photon.

So, to reconstruct entire trajectories, Steinberg's team performed a series of such measurements on a whole ensemble of identical photons as the photons passed from the double slit to the camera. These were weak measurements, so they were average values computed for a large number of photons. The measurements were repeated for different locations between the double slit and the camera, by placing the calcite crystal at varying distances from the double slit, each time in a plane parallel to the plane of the slits. These gave the average momenta of the particles in each plane.

There was one more measurement of relevance: the position of the photon as it crossed the plane of the calcite crystal. Optics were used to capture the image of each photon on the CCD camera as it passed through the crystal. This was used to calculate the positions of the photons. "The important thing is that we can't follow any given photon from plane to plane, but in each plane we can correlate position with direction," said Steinberg. So "in each plane, we construct a map of momentum versus position, and then connect the 'arrows' to build up trajectories."

The net result was that the researchers were able to reconstruct the average trajectories of the photons. The seemingly impossible had been done. The reconstructed trajectories looked very much like the simulated Bohmian trajectories. It has to be pointed out that one can arrive at the same predictions using standard quantum mechanics, so it's impossible to use the experiment to say which interpretation is correct. But despite similar predictions, the two interpretations have entirely different things to say about the nature of reality: in Bohmian mechanics, particles and their trajectories exist independent of observation, whereas in standard quantum mechanics the act of observation creates reality.

In 2011, *Physics World* named Steinberg's experiment the Breakthrough of the Year, and said, "The team is the first to track the average paths of single photons passing through a Young's double-slit experiment—something that Steinberg says physicists had been 'brainwashed' into thinking is impossible."

Not everyone is convinced that Steinberg's team has actually reconstructed Bohmian trajectories. One naysayer is Basil Hiley, who is now professor emeritus and still active. Hiley contends that the Bohmian trajectories shown in his 1979 paper with Dewdney and Philippidis are for nonrelativistic particles that are moving far slower than the speed of light. Photons are massless particles that move at the speed of light and so are relativistic. Particles of matter, such as atoms, are nonrelativistic. Hiley argues that Steinberg's experiment with photons is not the correct one to reconstruct particle trajectories—even though he's impressed with the experiment itself.

To verify the trajectories for nonrelativistic particles, Hiley has teamed up with Robert Flack and his team at University College London (UCL), and they are working on their own double-slit experiment. Hiley and Flack have chosen to work with argon atoms. The idea for their experiment came about a few years ago, when the two happened to meet over breakfast at a conference in Sweden. They began chatting about possibly testing Bohm's ideas, and Flack (an experimentalist) said, "I think we could have a go."

But it's been a hard slog. "Sometimes I wish I hadn't said that," joked Flack when I met him and Hiley in their lab in the basement of the physics department at UCL.

"Come on, Rob, you know, it's made your life worth living," Hiley bantered back.

The experiment involves first creating a cloud of argon atoms in an excited, metastable state (meaning they are stable in this state for about 100 seconds), and beam these atoms toward a double slit. After going through the two slits, and before they hit a detector, the atoms have to pass through a magnetic field. And just as in Steinberg's experiment, where the angle of the path of the photons through the calcite crystal is reflected in a change in the angle of polarization of the photons, the path the argon atoms take through the magnetic field is also reflected in certain internal properties of the atoms. This is a weak measurement. A series of such weak measurements can then be used to reconstruct the trajectories of the atoms. In theory.

The team is still working on getting the experiment up and running. I couldn't help but comment on how workmanlike these quantum physics experiments look. Optical benches, vacuum pumps, lasers, mirrors, and the like strewn about everywhere (the experimentalists would point out that everything is rather precisely positioned). The whole place looks like a mechanic's shop, albeit a very clean one with no grease, and it belies the profundity of what's being tested: the underpinnings of reality. "As a theoretician, I didn't think I'd live to see the day when the Bohm theory was being tested effectively," Hiley said.

Even though some, like Hiley, may disagree with Steinberg's experiment, the average trajectories measured by Steinberg's team brought out one aspect rather clearly: the paths of the photons did not cross the midline of the double-slit apparatus. The photons going through the left slit ended up on the left half of the far screen or camera, and those going through the right slit arrived at the right half of the detector. The paths would converge near the middle, but never cross.

While the trajectories were in line with what's expected from Bohmian mechanics, there was one nagging concern yet to be addressed. In 1992, Marlan Scully and his colleagues (dubbed ESSW, after the initials of the four team members), argued that Bohmian mechanics predicted something rather strange. If you could put a detector near the slits that could somehow tell which slit the particle went through without destroying the particle, ESSW showed that in some cases the detector near the left slit would fire, but the particle would end up at the right half of the far screen. According to Bohmian mechanics, a particle that hit the right half of the screen could have originated only in the right slit, because the trajectory cannot cross the midline. So, then, why was the math showing that sometimes, even though the particle hit the right half of the screen, the left slit detector fired? ESSW caustically called this out. "Tersely: Bohm trajectories are not realistic, they are surrealistic," they wrote.

"To them, this was a kind of *reductio ad absurdum* for the Bohm interpretation," Steinberg said. Over the years, many researchers (including Hiley) pointed out various problems with the ESSW analysis. Bohmian mechanics itself kept being tweaked, with physicists developing different versions of it, but the question of surreal trajectories never quite went away. "In essence, with any of these versions of the Bohm interpretation, you can find situations where one detector fires, but the Bohm model has the trajectories going through the other [slit]," said Steinberg. Surreal trajectories were a knock against the correctness of Bohm's ideas.

Steinberg's team stepped up their game, bringing more sophisticated experimental techniques to the optical bench. They wanted to know: could surreal trajectories destroy Bohm's ideas?

As an experimentalist, Steinberg is agnostic about theoretical interpretations of quantum mechanics. Even so, given the ESSW take on surreal trajectories, Steinberg was concerned about the validity of Bohmian mechanics. This despite the fact that the theory came with certain advantages. First, it restores determinism. "To a lot of people, the standard interpretation of quantum mechanics, which is very mathematical and abstract, gives up determinism, and they don't understand why you would give it up if you don't have to," said Steinberg. "They say that it's such an important philosophical assumption that if you showed me that it was in contradiction to reality, I'd give it up, but otherwise I'll bend over backwards to keep it." Bohmian mechanics keeps determinism intact.

Second, it makes nonlocality more explicit. Tests of Bell's inequality clearly show that the quantum world is nonlocal. "In the standard theory, the nonlocality looks mysterious, it looks like this spooky action at a distance, whereas in the Bohmian mechanics, it shows up exactly in the equations of motion," said Steinberg. It's clear how the motion of any given particle is instantly influenced by other particles. It's built into the mathematics.

Of course, Bohmian mechanics messes with the elegance of Schrödinger's ideas, in which there is just the quantum state of the system in question (given by the wavefunction) that evolves according to the Schrödinger equation. "You can think of it as being a particle or a wave, but it is what it is," said Steinberg. "In Bohmian mechanics, everything becomes two things. Everything is a particle *and* a wave. You have doubled the number of entities out there. That doesn't really disturb me, but it's an argument some people make."

What did disturb Steinberg, though, were surreal trajectories. "I also shared the ESSW intuition that these trajectories didn't make any sense," Steinberg told me. It "was one of the things that used to make me less enamored of the Bohmian picture." After their tour de force 2011 paper on average trajectories of photons going through a double slit, it was time to test the ESSW claim about surreal trajectories.

It required a small but significant tweak to their earlier experiment. Instead of a photon source that generated single photons, they began using a source that generated a pair of entangled photons. These photons are entangled in their polarization. The photons are polarized in the horizontal-vertical basis, so if one photon is measured and found to be polarized in the horizontal direction, then the other photon will be vertically polarized, and vice versa.

Let's call one of the entangled photons the system photon. This is sent through the same kind of setup that was used to measure average trajectories. The only difference being that instead of using a standard beam splitter, they used a polarizing beam splitter (PBS) to steer a vertically polarized photon into the left optical fiber and hence the left slit of the virtual double slit, and the horizontally polarized photon to the right slit. As we saw in earlier experiments, the polarization is converted into a path.

The other photon is the "probe" photon: it contains the information needed to probe which way the system photon went, without disturbing the system photon.

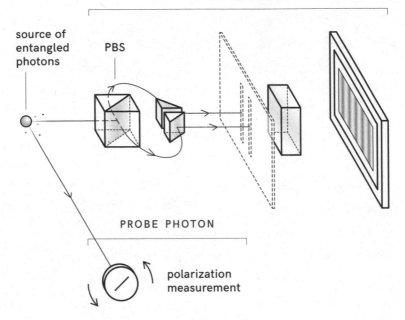

There are myriad things one can now do with this setup. For instance, if one simply measured the probe photon's polarization in the horizontal-vertical basis and got either horizontal or vertical as an answer, it immediately reveals which path its partner system photon took through the double slit. So, for all those probe photons that are measured in the horizontal-vertical basis, the corresponding system photons that go through the double slit don't show any interference, because we know which path they took and so they act like particles.

But if you measure the probe photon's polarization in the +45-degree direction, things change. The measurement involves sending the probe photon through a +45-degree polarizer: either it comes out (in which case, it's polarized at a +45-degree angle) or it doesn't. Crucially, the information about whether it was originally horizontally or vertically polarized is now erased. The math says that now it's equally likely to be horizontally or vertically polarized. Consequently, the corresponding system photon is also equally likely to be horizontally or vertically polarized, and so ends up in a superposition of going through the left and right slits. Taken together, such system photons show an interference pattern on the CCD camera.

This is essentially Marlan Scully's quantum eraser experiment.

Steinberg's team had lots more to do besides erase the which-way information. For starters, they began measuring the average trajectories of the system photons through the double slit. But for each weak measurement they made on a system photon, courtesy of the calcite crystal, they also measured the probe photon, to see if it was polarized at some given angle. What effect did this polarization measurement have on the system photon traveling through the double slit?

The team found that their choice of polarization angle for the measurement of the probe photon had an immediate effect on the system photon: its trajectory was altered (as determined via measurements over many, many particles). "So we directly see that this is a nonlocal theory," said Steinberg. "We can't really predict what these trajectories are without paying attention to the [probe] photon."

It was now finally time to ask the big question: were some trajectories surreal? To answer the question, they began studying the trajectories of the system photons, and for each trajectory, they looked at the polarization of the probe photon at various points in

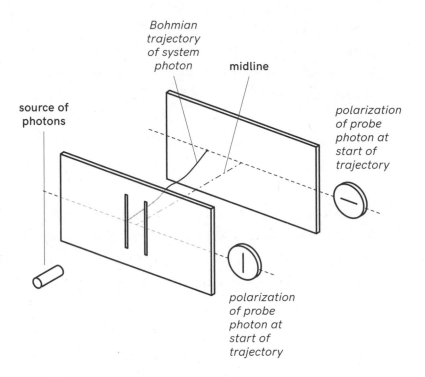

the trajectory of the system photon. Did the polarization change with the trajectory?

The answer was a clear yes. Say a system photon's trajectory started off at the left slit. The probe photon's polarization would be vertical. But as the system photon moved through the apparatus, the probe photon's polarization kept changing: another demonstration of the nonlocal interaction between the two photons. And there were situations when the system photon that began its journey at the left slit would reach the left half of the CCD camera screen, but the probe photon would end in a 50-50 mix of being horizontally and vertically polarized: its polarization was equally likely to be horizontal or vertical. More to the point, since the polarization of the probe photon is an indication of which slit the system photon took, the probe photon would sometimes indicate that the system photon took the left slit and at other times indicate it took the right slit, even though the Bohmian trajectory clearly showed that the system photon started at the left slit and remained in the left half of the apparatus, never crossing the midline.

This was the surreal trajectory ESSW had theoretically identified. Except, to them, it had made no sense, because the Bohmian trajectories were at odds with the information in the which-way detector. But Steinberg's experiment shows that the system photon, as it's moving from the slit to the screen, is nonlocally influencing the polarization state of the probe photon: it's nonlocally influencing the which-way detector. So, sometimes, at the end of the system photon's trajectory, the probe photon's polarization may have changed, say, from horizontal to vertical, leading one to erroneously interpret that the system photon went through the right slit, not the left one. If you didn't know of the nonlocal interaction between the photons, you would see the results as a knock against Bohmian mechanics, as ESSW did. Steinberg's team is arguing that Bohmian mechanics is consistent and defensible, and can't be ruled out based on the ESSW argument.

The probe photon shows the correct polarization value when the system photon is near the slits, but sometimes, at the moment the system photon hits the CCD camera, the probe photon's polarization itself has changed, due to nonlocality. So the trajectory is surreal or untenable only if you think the which-way detector's final value was always its value. That's clearly not the case.

While the findings remain controversial (partly because there are a few different versions of Bohm's theory and partly because of the debate over the meaning of weak measurements), for some, including Steinberg, the fact that surreal trajectories have a perfectly sensible explanation makes it possible to think of Bohmian mechanics as a viable alternative to the Copenhagen interpretation. The experiment shows that it cannot be ruled out, yet.

Of course, the same experimental results can be predicted using standard quantum theory. "It certainly doesn't come down on one side or the other," said Steinberg. "The best thing experiments like this can do is to remind people that the [Bohmian] interpretation exists—people who had either forgotten or never knew about it— and show them that although it might sound mysterious when you first hear it described, these hidden trajectories are related in a very straightforward way to things you can easily imagine going into a lab and measuring."

It took years for Bohm to get out of exile from Brazil and reach the UK, but it's taken even more time for his ideas to start being treated as valid and worthy of such experimentation. "It's an interpretation that hasn't gotten enough attention, people aren't aware of it, and we want . . . to bring it back to its rightful place among all other interpretations," Steinberg said.

Philosopher David Albert thinks the cold shoulder Bohm's ideas received was in no small measure due to the politics of the time. "A large part of the story of the reception of Bohm's theory has to do with the fact that when he was in the middle of proposing this he refused to testify before the Un-American Activities Committee and was hounded out of the country. A lot of the reception of Bohm's theory is tied up with that," said Albert. "Science is a very human endeavor, and the history of foundations of quantum mechanics are a particularly vivid example of this."

Even the rise of the Copenhagen interpretation, Albert argued, could be seen in the light of the "crisis of representation" that engulfed literature in the late 1800s and early 1900s. Could language capture objective reality? Modernist literature said no: it played with perspective, highlighting uncertainties and ambiguities inherent in any one view of the world. It was "inspired by the modern realization of the observer's role in both creating and curtailing the world of perception" and led to, at its extreme, "the view that there is no true world, since everything is but 'a perspectival appearance whose origin lies in us.'"

"Literary modernism was supposed to be a response to the crisis of representation in literature. Physics wanted to have its crisis of representation too," said Albert. And it got one, when quantum physics claimed that "there isn't any such thing as telling the flat-footed, objective, true story of what's happening to the particle in between this measurement and that measurement."

Bohm's ideas certainly challenge this view. Goldstein is more than aware that the rumblings about Bohm's ideas are growing. "After decades and decades, people are taking Bohmian mechanics a little bit more seriously," he told me. "There was a time when you couldn't even talk about it, because it was heretical, it wasn't Copenhagen. There was a kind of political correctness about physics. It probably still is the kiss of death for a physics career to be actually working on Bohm, but maybe that's changing."

But as much as Bohm's realist, nonlocal theory appeals to some, there are others for whom it holds less sway. Even Steinberg, whose experiments are responsible for shining a favorable light on Bohmian mechanics, remains quietly skeptical of Bohm's ideas. One problem for Steinberg is that Bohmian mechanics privileges position over other properties of a particle: it's only position that gets the honor of being associated with a hidden variable. But what about a particle's spin or its polarization? The theory treats these differently, and doesn't accord them the same kind of hidden variable as it does position. "I must admit I always found that distasteful," said Steinberg. "Because by the time I was raised in quantum mechanics, I didn't see anything that special about position. There should be one consistent treatment of all kinds of measurements. If your measuring device uses polarization, there should be a different hidden variable that corresponds to polarization."

Goldstein thinks otherwise about according position the pride of place as a hidden variable. "I regard it as very much a virtue that one does not need to have additional reified observables in order to fully understand what is going on in any quantum measurement," he said. "Position suffices. People have indeed proposed adding Bohmian versions of the other observables. The result is ugly and the effort rather pointless."

While Goldstein strongly favors Bohmian mechanics, Steinberg's holding out for something more than either orthodox quantum mechanics or Bohmian mechanics. "I'm most attracted to the possibility of us discovering something beyond quantum mechanics that resolves these problems by saying that this wasn't the complete theory to begin with," he told me. "I'm waiting."

Roger Penrose, a theoretical physicist at Oxford University in the UK, shares Steinberg's view that quantum mechanics may be incomplete. "Quantum mechanics is a provisional theory," he told me when I met him at his home. Sitting in his bucolic backyard on the outskirts of Oxford, Penrose proceeded to explain why gravity (which we have ignored thus far in our attempts at understanding the quantum world) may have something to do with fixing quantum mechanics (at least in the minds of those who think it needs fixing). And as always, it began with a discussion of the double slit. This time, however, instead of a particle going through two slits, Penrose talked of a cat walking through two doors at once.

7

GRAVITY KILLS
THE QUANTUM CAT?

The Case for Adding Spacetime into the Mix

A university student attending lectures on general relativity
in the morning and . . . on quantum mechanics in the afternoon, might be forgiven for concluding that his professors
are fools, or they haven't talked to each other for at least a
century.

— **Carlo Rovelli**

The day I'm supposed to meet Roger Penrose at Oxford University, he has to stay home. So I receive instructions to get to his
house—a map that he drew by hand almost two decades ago for
guests coming to his housewarming party, and which he has been
updating ever since, as the neighborhood changes around him. The
high-level map of where he lives near Oxford is at a different scale,
but is informative enough, and there's even a detailed, zoomed-in
drawing of the roads and houses within a few hundred meters of his
home (from macroscale to microscale, I think to myself). It contains

warnings: "Fancy blocked entrance (not us!)," "Imposing old house, NOT us," "Smart blocked entrance. NOT US." And an arrow pointing to his house, saying, "Our broken down entrance."

Penrose's penchant for drawing by hand is obvious when he gives talks. Eschewing sophisticated presentations with graphics and animations, he still relies on an overhead projector, oftentimes using multiple hand-drawn transparencies to project a single image, each transparency meticulously drawn and annotated in different colors. He then carefully fiddles around with them, stacking them, sliding them around, until a complex story emerges—like a cat in a superposition of being alive and dead at the same time.

A mathematical physicist, Penrose is best known for his work on general relativity and cosmology, particularly on singularities that occur inside black holes or at the big bang, the places in the cosmos where our known laws of physics break down. Such breakdowns happen partly because the physics of intense gravitational fields has to coexist with the physics of the microscale. General relativity, which is a theory of gravity on large scales, has to confront quantum mechanics. But so far, gravity has stubbornly resisted succumbing to quantization, unlike the other three fundamental forces of nature (the quantum of the electromagnetic force is the photon, but there is no observed quantum of the gravitational force). There's a consensus view, however, among those striving for a theory of quantum gravity that it's general relativity that has to make way in this tussle for supremacy, and that quantum mechanics can continue untouched for the most part.

Penrose, however, thinks differently. "There has to be give on both sides," he told me. "It's not as though one wins over the other;

it's got to be an even-handed marriage." And that union, said Penrose, means fixing what he thinks is wrong with quantum mechanics. "The trouble with quantum mechanics [is that] . . . it doesn't really make sense," he said.

He paused before continuing. "I shouldn't appeal to authority here, you see," he said. "You've got authorities on both sides." Still, he pointed out that Einstein, Schrödinger, de Broglie, and even Dirac to an extent had all felt that something was not quite right with quantum mechanics. Schrödinger magnified his unease with his eponymous cat—a thought experiment that's a clear affront to our classical sensibilities.

To illustrate the absurdity, Penrose has his own version of Schrödinger's cat, a "more humane version," he quipped. The cat is in one room and is confronted with two doors that lead to another room. The mechanism that opens one or the other door is quantum mechanical. Penrose imagines a photon going through a beam splitter—if it's reflected it opens the left door; if it's transmitted it opens the right door. This results in the system being in a superposition of "left door open, right door closed" *and* "right door open, left door closed." If the cat goes through either door, it gets some food, but unlike the double-slit experiment, where the particle enters a superposition of going through both slits simultaneously, our classical sensibility says that the cat cannot go through both doors at once. But "quantum mechanically, you'd have to consider that both alternatives coexist in order to get the right answer," said Penrose.

Treating the cat quantum mechanically leads to the wavefunction of the cat going through both doors in a kind of superposition of motions. According to the Copenhagen interpretation, some interaction with a classical system that can be considered an act of measurement, such as a CCTV camera recording the cat's entry, would then collapse the wavefunction and show the cat going through one or the other door. As with most physicists who have trouble with quantum mechanics, Penrose finds the idea that measurement is necessary to collapse the wavefunction implausible.

One way out of this mess is if there is a clear divide between the quantum and the classical—and so the cat is always a classical object and cannot be treated quantum mechanically. Penrose has had a radical idea for decades that such a divide exists and it comes about because of the spontaneous collapse of the wavefunction without necessitating a measurement, explaining why an object as large as a superposed cat would remain in a superposition for only a small, small fraction of a second before collapsing into one classical state. In the case of Schrödinger's cat, Penrose's theory could cause a collapse of the total system, so the cat would be either dead *or* alive almost instantly.

The solution involves gravity and makes rough predictions about where we might find the classical-quantum boundary. "You have got to look not at the impact that quantum mechanics has on gravity, but the impact gravity has on quantum mechanics," he said.

On that rather nippy English afternoon, sitting at a wooden table on a deck in his backyard, Penrose took off his glasses and placed them on the table. Glasses have mass, and according to general relativity, they will warp or curve spacetime in their vicinity. Gravity is the curvature of spacetime: the more massive the object, the greater the curvature (black holes really put a dent in spacetime, a pair of glasses, not so much). But if the glasses were in a superposition of being in two places—Penrose moved them back and forth for his show-and-tell—then the glasses at one location would warp spacetime one way, and another way at the second location. "Now, therefore, you have a superposition [of] two slightly different spacetimes," he said. And that, said Penrose, is an unstable situation that destroys the superposition rapidly if the mass displacement is large.

Say you have an experiment in which you have put a little lump of material in a superposition of being in two different places, said Penrose. "I'd claim that the superposition will spontaneously become one or the other in a timescale which you can calculate, roughly."

According to Penrose, a superposition of two spacetimes creates what he calls a "blister" in the four-dimensional volume of spacetime. When that blister grows to one Planck unit in four dimensions, where three of those are space dimensions (one Planck length equals about 10^{-35} meters) and one is the dimension of time (one Planck time is of the order of 10^{-43} seconds), the superposition will spontaneously collapse to one state or the other.

For his pair of glasses, such a blister in spacetime would form in far less than one Planck time. "It'd be instantaneous," said Penrose. That is why we never see macroscopic objects in a superposition of states, whereas for subatomic particles, such a spacetime blister would take practically forever to resolve itself into one side or the other. "It could be about the age of the universe."

Penrose has another way of thinking what happens when two spacetime configurations are in superposition. Take his glasses again. They have a property called gravitational self-energy, which is the energy required to hold the system together if there were no other forces, in a configuration resembling a pair of glasses. This is, of course, true of all matter, not just Penrose's glasses. Now, if the glasses are in a superposition of being in two places, then there's an uncertainty about the gravitational self-energy of the system. Penrose then resorts to one of Heisenberg's uncertainty relations, to do with simultaneous measurements of energy and an interval of time: the more precisely you know the energy of a system, the less sure you are of the time interval, and vice versa. By applying this uncertainty relation to the uncertainty in the gravitational self-energy of the system in superposition, Penrose estimates the time interval for which the superposition can be stable before collapsing to one or the other state. Admitting to a bit of "hand-waving," Penrose said, "I can't say when it'll happen and I can't say which it'll do, but I can give you an estimate."

As convinced as Penrose is of his idea that gravity must play a role in the collapse of the wavefunction, the response from physicists and philosophers probing the foundations of quantum mechanics has been tepid. It's probably because he's proposing modifying quantum mechanics—particularly the way the wavefunction evolves according to the rules of the Schrödinger equation. The gravitationally induced collapse of the wavefunction messes up this rather beautiful picture. But then it does provide an explanation for why there exists a boundary between the quantum and the classical.

John Bell always found this boundary, unspecified but implicit in the Copenhagen interpretation, very troubling. After all, even a so-called classical measurement apparatus consists of atoms and molecules, each of which on its own is regarded as quantum mechanical, but an agglomeration of an unspecified number of atoms and molecules has to be at some point regarded as a classical object. Bell objected to what he called a "shifty split" between the two worlds. Penrose's work, while it messes with the elegant evolution of the wavefunction, gives a reason for why this split might exist in the first place.

Penrose is puzzled by people's objections to modifying quantum mechanics. He points out that Newtonian mechanics lasted a lot longer than quantum mechanics has. "People were pretty convinced that that was a picture that was going to stick forever," said Penrose. Yet it didn't. And that too despite not suffering from any measurement paradox. "So I don't see why people are so completely convinced by [quantum mechanics]."

Gravity induced collapse of the wavefunction can be seen as one example of a more generalized solution to the measurement problem. In 1986, around the same time that Penrose and others, including most prominently the Hungarian physicist Lajos Diósi (who came to these ideas a touch earlier than Penrose), were formulating ideas about gravity's role, three physicists, Giancarlo Ghirardi, Alberto Rimini, and Tullio Weber, came up with another way of modifying quantum mechanics.

GRW, as the theory is called, changes the way the wavefunctions of particles evolve. Rather than being completely governed by the Schrödinger equation, GRW adds a component to the dynamics of the wavefunction that causes it to collapse at random. But the collapse is not induced by gravity, à la Diósi-Penrose, or by a measurement, as in the Copenhagen interpretation. Rather, it is something spontaneous, and an elemental aspect of nature. It causes the wavefunction of a particle to go from being spread out to being relatively localized. Mathematically, the diffuse wavefunction, which says that the particle can be in many different locations at once, is multiplied by another function. Think of this function as something that is mostly zero at all physical locations but at one location rapidly rises to a certain peak. The net result of this multiplication is to collapse the wavefunction, leaving the particle roughly localized at one point in space and time.

To mimic the predictions of quantum mechanics, GRW has to ensure two things. One, that such spontaneous collapses are extremely rare for individual particles, so that they can remain in superposition of states for any measurable length of time. Two, that for a large collection of particles, say, those that make up a cat, the collapse of the wavefunction is near certain, so that the cat is always found in some macroscopically identifiable state and not in a superposition. In their earliest versions of the theory, GRW showed how to set up the theory so that it could take almost 100 million years for a single particle to collapse, whereas a macroscopic object with about 10^{20} particles would collapse almost instantly (a few tens of nanoseconds or less, but estimates vary).

As always with modifications to quantum mechanics, people found flaws with the GRW model, which others tried fixing with further tweaks. For example, the GRW model doesn't deal well with a large collection of particles with identical properties (say a bunch of electrons, which cannot be distinguished from one another). Another version of spontaneous collapse fixes this problem. And of course, the parameters of the GRW model can themselves be fine-tuned in an ad hoc manner to suit the outcome of experiments—something that bothers naysayers. Nonetheless, the basic idea behind all these models is still the same: a spontaneous collapse of the wavefunction that has nothing whatsoever to do with measurement. "The collapse is something that is occurring all the time to every particle, at random, with a certain fixed probability per unit time," said philosopher David Albert, who has a soft spot for the GRW theory. "There is no need to talk about measurements, or anything like that. There is no need to use any of these words."

Even John Bell was suitably impressed when he first encountered the theory. "Any embarrassing macroscopic ambiguity in the usual theory is only momentary in the GRW theory. The cat is *not* both dead and alive for more than a split second," he wrote.

More important for Bell, collapse theories had the mathematics to back up claims of collapse. He said they "have a certain kind of goodness . . . They are honest attempts to replace the woolly words by real mathematical equations—equations which you don't have to talk away—equations which you simply calculate with and take the results seriously."

Some experimentalists are doing exactly that. Whether it's Penrose's theory or GRW-like theories, they all make potentially testable predictions about where the boundary between the quantum and the classical might lie. And even though the predicted boundary seems out of reach of today's experiments, it's not stopping Markus Arndt, an experimentalist in Vienna, from looking for it. He's doing that by sending larger and larger molecules (not just photons or electrons or atoms) through some complicated versions of the double slit and making them interfere. The day he can conclusively claim that molecules of a certain size don't interfere because of their size—meaning they can't remain in a coherent superposition of simultaneously taking two paths—he will have found nature's dividing line. For now, he's happy claiming that his team has worked with the biggest "Schrödinger's cat" ever to confront two doors at once.

Schrödinger's cat has become code for a macroscopic object that can remain in a superposition of multiple states. For Arndt, the molecules he works with are such objects. While they are certainly nowhere near as big as even the smallest possible cat, with the mass of 10,000 protons, they are the largest macroscopic objects that have thus far been seen in superposition going through a double slit. "I am claiming that we have the biggest Schrödinger cat," Arndt told me, tongue in cheek. To count as Schrödinger's cat, a quantum system should be "something that should be really macroscopic; it should be at least as warm as a cat and it should contain a biomolecule." The objects that Arndt has put into superposition are certainly macroscopic, and they contain biomolecules. But unlike a cat that's at room temperature, for experimental reasons, Arndt's molecules are much hotter. "A real cat would be dead by then," he quipped.

Arndt's interest in such experiments began when he was a postdoc with Jean Dalibard at the École Normale Supérieure in Paris (Dalibard, as a student in the 1980s, had worked with Alain Aspect on tests of Bell's inequality, but then did seminal work on his own, particularly on trapping atoms in place using lasers and magnetic fields). The Paris team demonstrated the validity of de Broglie's ideas of matter-wave duality, using cesium atoms.

Arndt continued his postdoc phase with Anton Zeilinger, first in Innsbruck, Austria, and then moved with Zeilinger to the University of Vienna, and now runs his own labs on Boltzmanngasse. Among the many experiments Arndt's group is doing, one with particular relevance to questions about the foundations of quantum mechanics has to do with molecular interferometry: doing advanced variants of the double-slit experiment with large molecules and nanoparticles. Zeilinger, in his early days, had been part of a team that demonstrated a double-slit experiment with single neutrons, the most massive particle to be tested in the 1970s. Soon physicists began showing that atoms could be placed in superposition of states and made to interfere. In 1991, Jürgen Mlynek and colleagues in Konstanz, Germany, sent helium atoms through two 1-micrometer-wide slits, about 8 micrometers apart, and saw the atoms interfere. (The history of atom interferometry is rich. Other prominent names include David Pritchard at MIT, who showed in 1983 that atoms could be diffracted at gratings, and Fujio Shimizu at the University of Tokyo, who reported in 1992 a double-slit experiment done with neon atoms.) Since then, it has become a race of sorts, with the attention shifting to molecules, to see who can bell the biggest Schrödinger's cat.

The key issue with getting molecules to interfere is to ensure that they don't hit any stray particles during the experiment. If a molecule interacts with a photon, or an electron, or an air molecule, the molecule being tested gets entangled with the environment. What starts off in a state of coherence undergoes decoherence. "In such a setting, I cannot detect [the which-way information], but the environment can," said Arndt. In principle, the mere presence of which-way information in the environment is enough to make a mess of the superposition. The best way to avoid decoherence, then, is to carry out the entire experiment in a vacuum chamber.

In 1999, Zeilinger, Arndt, and their team were the first to do a multi-slit experiment with large molecules consisting of sixty carbon atoms each—a stable form of carbon that had been identified in 1985 and named buckminsterfullerene, or buckyball, because it has the segmented 3-D shape of the geodesic dome invented by Buckminster Fuller. A buckyball is about 1 nanometer in diameter, and when flying in a molecular beam at a speed of 200 meters/second, it has—according to de Broglie's equation relating the wavelength of a particle to its momentum—a wavelength that's about 350 times smaller than the size of the molecule. As objects get more massive, their de Broglie wavelengths get smaller and smaller, and it's one of the reasons why we don't observe the wave nature of such objects in ordinary encounters. However, according to quantum mechanics, their wave nature should be apparent when they go through a double slit, one molecule at a time. The researchers showed that the C60 molecules could indeed be put into a superposition of taking two paths at once and made to interfere.

As with the experiment with single photons, Arndt is quick to stress that the interference being observed in these experiments is a quantum mechanical effect at the level of single molecules. A molecule can be described by a wavefunction for its center of mass. The amplitude of the wavefunction at different points in space lets us calculate the probability of finding the molecule at those locations. Each molecule that goes through the double slit has to have a wavefunction similar to the other molecules to ensure that the interference pattern that develops over time adds up; otherwise the pattern can get fuzzy or not form at all.

Getting photons or electrons or neutrons to have similar wavefunctions is relatively easy. Not so for molecules: you have to get them all moving in the same direction at the same velocity. A daunting task, since unlike atoms of gas, "molecules don't like to fly," said Arndt. They are more likely to stick to surfaces, to each other, do anything but go from the source to the double slit and beyond.

To get them to leave their sources, the molecules have to be heated or otherwise launched, but not in ways such that their internal thermal energy makes interference impossible. All of which, as the team has gone to bigger molecules, has meant resorting to some serious chemistry and designing custom molecules that have stable internal bonds between the atoms that make up each molecule, and yet the molecules are not drawn to each other, they are not "sticky." The team's best effort so far, in terms of molecules going through a multi-slit arrangement, is a whopper: it's a bespoke molecule with 284 carbon atoms, 190 hydrogen atoms, 320 fluorine atoms, 4 of nitrogen, and 12 sulfur atoms. That's 810 atoms in one molecule with a total atomic weight of 10,123. The molecule was synthesized by a team led by Marcel Mayor, of the University of Basel, in Switzerland. The high fluorine content acts like a Teflon shell—preventing the molecules from sticking to each other too readily.

When these molecules leave the source, they have a temperature of about 220 Celsius—which would kill a cat.

Cats or molecules, the task is to put the wavefunction of the macroscopic object into a superposition of states. In this case, to first get each molecule's wavefunction on the same page, the beam has to be collimated in both the horizontal and vertical directions, which simply involves letting the molecules pass through narrow openings and selecting only about the one in ten million that make it through. Now, the molecules are moving in a narrow beam, but they still might have very different velocities, and hence different wavefunctions. So the molecules are further filtered—they have to go through three narrow slits that are placed at different distances and heights, such that a trajectory that passes through these three slits traces a parabola. Imagine throwing a ball. It'll travel in a parabolic arc, the shape of which will depend on the speed at which you throw the ball. Or inversely, for any given parabolic arc, all balls with that trajectory have the same speed. Arndt and colleagues took advantage of this simple fact. They arranged the three slits so that all the molecules that get through them would follow the same parabolic path, and so have a definite velocity when they exit. Now they had a collimated beam of molecules with similar velocities (within about 10 to 15 percent of one another)—and hence similar wavefunctions.

There was another big hurdle to cross to get these molecules to interfere. For a particle to be in a superposition of going through both paths, its wavefunction has to spread out enough to span two slits. For the kinds of distances traveled on laboratory benches, this is not an issue for particles such as photons and electrons. But not so for a molecule: it'd have to travel a long distance for its wavefunction to spread out enough, making for an impossible experiment. So Arndt's team deployed a trick. They made the molecules first confront an array of extremely narrow single slits. This causes diffraction at each of these slits, and the wavefunction starts spreading rapidly on the other side of each slit. Now when the wave front reaches a grating with multiple slits not too far away, it's wide enough to encounter at least two slits at the same time and enter into a superposition. To get a sense for just how small the dimensions involved are, the two slits are only 266 nanometers across (about a hundred thousand times smaller than the width of a human hair). To ensure that the molecules have a chance of hitting the slits (there's no way to steer them precisely to any one location), the team illuminates a multi-slit grating with a molecular beam just one millimeter across that spans about 4,000 slits. Any one molecule's wavefunction will hit only two adjacent slits out of these 4,000—so effectively, each molecule sees only a double slit.

One final challenge remained: detecting where the molecules land after crossing the double slit. With a photon, this is relatively easy. A photographic plate can register a hit. Molecules are lumbering beasts compared to photons. "If they [land] onto a surface, they start rolling around, and if they do that, they smear the interference pattern," said Arndt. "So you have to make sure that the molecules are bound, wherever they hit the surface."

This meant designing special surfaces to capture molecules and make them stick the landing. One solution the team came up with was something called reconstructed silicon, which is essentially ultra-pure silicon with naked chemical bonds on the surface, like so many arms waiting to entrap molecules. The giant molecules land on the silicon surface and bond. Over time, these molecules pile up at various locations along the silicon screen.

But unlike a photographic plate, the pattern made by the molecules is not visible to the naked eye. Arndt's team had to study the surface using a scanning electron microscope—and what they saw was an interference pattern. Molecules amassed in areas that made up bright fringes, with fewer in locations that made up the dark fringes.

It's worth reiterating that the molecules are not interfering with each other. This is single molecule interference: in the language of standard quantum mechanics, each molecule ends up in a superposition of going through two slits at the same time, and these two states interfere, causing the molecule to go to locations that end up as bright fringes and avoid places that become dark fringes.

"This is the most pictorial representation of the weirdness of quantum physics, that you can see things [behaving] as if they were in various places at the same time," said Arndt. "Of course, it gets more and more counterintuitive, at least psychologically, if things become bigger and bigger and more complex internally. It relates to the question: why can *I* not be in two places at the same time?"

The ambiguous language that's needed to talk about what's happening is not surprising. The molecules are particles, individual "things," and yet the experiment has to acknowledge not only their de Broglie wavelengths but also each molecule's wavefunction, and the spreading out of the wavefunction. Treating molecules as real particles, with real trajectories, while a wavefunction is going through both slits, has Bohmian overtones.

"To be honest, if you are looking at [these] matter waves, occasionally you think as a Bohmian. It's very hard to avoid," Arndt said. "When we describe our interferometers, we are always thinking of the entirety of the particle, its mass, electrical properties, and internal dynamics, etc. Whenever it interacts with a grating, it's always there as an entire particle and yet it must have had information about several slits, somehow. In this context, it's more intuitive to think there's a pilot wave driving the particle around. That fits most nicely with Bohm's theory."

But thinking in Bohmian terms is not the Viennese way. "Vienna does not have a tradition of appreciating the role of the de Broglie-Bohm mechanics," Arndt told me. No wonder, given that the patriarch of the Viennese school of quantum mechanics, Anton Zeilinger, is a strong non-realist in the tradition of Niels Bohr and the Copenhagen school of thought.

Arndt is quick to point out that despite his tendency to think in Bohmian terms for matter waves, he's a non-realist when it comes to the dynamics of the internal states of molecules, in that these states don't exist until we measure them. Besides, what he's really after is to discern if there is a quantum-classical boundary that's predicted either by Penrose's gravitational collapse theory or by any of the many flavors of the GRW collapse theory. Neither prediction, unfortunately, is within easy reach. Arndt recalled the early days of the GRW theory, when it was thought that molecular interferometers would see collapse and hence find the boundary between the quantum and the classical, with molecules at about 10^9, or a billion, atomic mass units. Experimentalists could dream of testing the theory. Subsequently, some theorists revised the target to about 10^{16} atomic mass units, making the theory extremely difficult to falsify. "Theorists have a simple life," said Arndt. "They can change their parameters."

Life isn't as simple for experimentalists. If the molecules get bigger, they have to be made to move slower; otherwise their de Broglie wavelength will get so small that it'd be impossible to make fine-enough slits to see interference. And even if they could figure out how to make molecules fly slowly, there's yet another problem to confront. Slower molecules will take longer to get through the double-slit apparatus, and when the molecules are in flight for longer, Earth's rotation starts becoming an issue. The molecules are flying straight through the vacuum, decoupled from everything, and if their time of flight is more than a few seconds, the vacuum chamber and the gratings would have moved enough due to Earth's rotation that the entire experiment goes out of alignment.

Arndt's calculations show that they can experiment in their lab with molecules of up to 10^8 atomic mass units, which is about 10,000 times more than the current record. They are even experimenting with sources of biological samples, like the tobacco mosaic virus, which has a size of about 10^7 atomic mass units. That's "in the range of what you could do in our lab, hypothetically," said Arndt. But the viruses haven't cooperated. "We had many different efforts to launch them and each time they broke apart."

One way to go to bigger masses is by using metal or silicon nanoparticles. Experiments with larger particles would need to negate the effect of Earth's gravity and hence would have to be done either in space (a very expensive proposition) or more likely in a drop tower, a special tower from which enough air has been pumped out to let objects free-fall without any air resistance, effectively mimicking conditions in outer space for a few seconds. There's a 146-meter-high tower in Bremen, Germany, built specially for such experiments. In principle, a fully sealed vacuum chamber that encapsulates the double-slit experiment can be dropped down the tower, and during free fall—for about 4 seconds—the molecules and the experiment won't experience Earth's gravity, and thus everything would remain in alignment.

While verifying collapse theories with such experiments remains a distant dream, Arndt hasn't ruled out seeing modifications of the established evolution of quantum systems at mass scales smaller than predicted by either Penrose's or GRW-like theories. "The experimentalist in me says, 'Well, who knows?' The models are made up by clever people and nobody knows whether they are true. It may well be that something happens well before. No one knows. So one should just do the experiments, see what happens."

If nothing happens—meaning molecules remain in superposition and their coherence is preserved—then it's telling us that there is no quantum-classical boundary, at least at the mass scales being probed by the experiment. On the other hand, "if [coherence] is not preserved, it's a major discovery," says Arndt. "In either case, you win."

It may prove impossible to do the classical double-slit experiment with molecules large enough to find evidence for collapse, especially at the mass ranges predicted by Penrose's theory. But what if an element of the experimental apparatus itself can be put into a superposition of states? Besides having a photon go through one arm or the other of a Mach-Zehnder interferometer, what if one of the mirrors used in the interferometer could be made small enough to be put into a superposition of being at one place or another. It's analogous to putting one of the slits in the double-slit experiment—which we have until now considered to be a macroscopic, classical, and immovable part of the apparatus—into a superposition of two positions. This would have a very strange impact on the photon going through the interferometer. Not only is it confronting two slits at once, but it's as if one of those slits is itself in two different positions. It turns out that this type of interferometer is ideally suited to testing Penrose's collapse theory.

Dirk Bouwmeester, a Dutch experimentalist, has been working on one such experiment for more than a decade, an idea first suggested to him by Penrose himself. When Bouwmeester was working on his PhD in the Netherlands, he became interested in certain solutions of Maxwell's equations of electromagnetism, in which light goes around in knots. He realized that what he was studying was closely tied to Penrose's work on twistor theory (one of Penrose's signature contributions to theoretical physics). In twistor theory, the most fundamental things in nature are not particles but rays of light, or twistors. Bouwmeester was still a student when Penrose came to the Netherlands to give a talk. Bouwmeester nabbed him afterward to discuss twistors. Penrose was getting ready to leave and suggested Bouwmeester come with him to the airport and they talk on the way. As it happened, "it was terribly bad weather and the flight was delayed, and we ended up talking a bit longer. That was the first time I met him," Bouwmeester told me.

That interaction led Bouwmeester to apply for a postdoc at Oxford University. After a year studying twistor theory at Oxford, Bouwmeester moved to Innsbruck, Austria, to work with Zeilinger on quantum teleportation and entanglement. With that experience in hand, Bouwmeester came back to Oxford to set up his own quantum optics laboratory. It was during this second stint at Oxford that Penrose walked into his lab one day and said: "I have this experiment that we need to do."

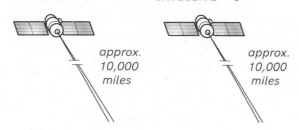

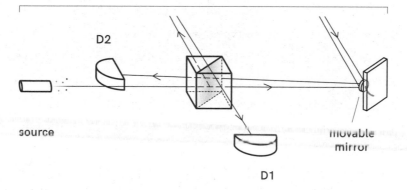

It was the strangest thing. Penrose had plans for doing an interferometry experiment in space, which involved three satellites. It goes something like this. On one satellite, "A," you first send an X-ray photon through a beam splitter. The photon ends up in a superposition of being reflected and transmitted. The reflected photon is sent on its way to another satellite, "B," about ten thousand miles away. The transmitted photon, which is still on satellite A, goes toward a tiny mirror. The mirror is attached to a cantilever, so that it can move if something hits it. The mirror is so small, and the X-ray photon has so much energy, that even as the photon impacts the mirror and is reflected at right angles, it displaces the mirror a smidgen. This photon too is now sent off toward another satellite, "C," which is a similar distance away from A as is B (note that technically it'd be quite easy to have satellites B and C combined into one satellite; it'd make the experiment cheaper too).

Quantum mechanics says that the photon is in a superposition of taking two paths. Not just that, the tiny mirror is also in a superposition—of being displaced or not-displaced, of being in two locations, barely about 10^{-13} meters apart, about halfway between the size of an atomic nucleus and the atom itself (the precise displacement depends on the type of mirror and cantilever).

The photon paths reach the two satellites, each of which has a rigid mirror that bounces the photon right back to the first satellite. The photon bounced off satellite B comes back to the beam splitter. The photon bounced off satellite C, before it can reach the beam splitter, has to again encounter the mirror that it had previously displaced. The distance between the satellites and the stiffness of the cantilever is set such that the moving mirror is back to its original position at exactly the same time the returning photon encounters it. The momentum from the mirror is transferred back to the photon, reflecting it at right angles toward the beam splitter. The mirror goes back to being at rest.

The two photon paths are designed such that the reflections from satellites B and C both reach the beam splitter at the exact same instant. So if the photon is still in a superposition of having taken both paths, then the two paths will constructively interfere, and the photon will exit the beam splitter toward detector D1. Crucially, the photon will never exit toward detector D2, because that direction represents destructive interference.

If this is awfully reminiscent of interference in a Mach-Zehnder interferometer, that intuition is not off the mark. This is yet another way to make two paths that light can take interfere: this particular arrangement is called a Michelson interferometer (with a tiny Penrose variation thrown in—the movable mirror).

Okay, so why go to all this trouble? Why all the fuss about using satellites in space? For one, the vacuum of space ensures that there is very little chance of the photon or the mirror hitting stray particles, an interaction that will lead to decoherence and loss of superposition. Also, the vast distances between the satellites ensure that the photon remains in superposition for a long enough time. This long so-called coherence time is necessary to test Penrose's ideas.

According to Penrose, the size of the tiny, movable mirror makes all the difference as to whether or not the photon ever goes to detector D2. When the photon is in a superposition of heading toward satellites B and C, the mirror is in a superposition of being displaced and not-displaced. Penrose's gravitational collapse theory says that the higher the mass of the mirror, the faster it'll collapse into one position or the other.

Let's say that the collapse never happens in the time it takes for the photon to return to the beam splitter and hit one of the detectors. In that case, the photon will come back in a coherent superposition of having taken two paths and hit detector D1.

But if the mirror's quantum state were to collapse before the photon reaches the detectors, then the photon will also collapse into having taken one path or the other. That's because the wavefunctions of the mirror and the photon are entangled, and their fates are tied up with each other. If such a collapse happens when the photon is still en route, then the photon will arrive at the beam splitter having taken one *or* the other path, not in a superposition of having taken both paths. It acts like a particle. There is no interference. So the photon has an equal chance of going to D1 or D2.

For a given mass of the mirror, if you did this experiment a million times, and the photon always went to D1, then you can say that the mirror never collapsed. But if half the time the photon ended up at D2, the mirror has collapsed in each run of the experiment. This becomes a way to verify Penrose's ideas of the collapse of macroscopic objects due to gravity—and find the mass scale at which collapse happens.

When Penrose walked into Bouwmeester's lab, he was keen on actually doing this experiment in space. He knew people at NASA; he thought they could pull it off. Bouwmeester had to bring the discussion back down to Earth. "My initial reaction was—it's a very interesting problem to work on, but my expertise is in optics," Bouwmeester told me. "Let's see if we can design something that fits on an optical table."

And they came up with a solution, with help from a talented postdoc named Christoph Simon and a gifted PhD student, William Marshall. Sitting in his office at the University of California, Santa Barbara, in 2017, Bouwmeester showed me a picture taken in 2001 of the four of them standing in front of an optical bench. "That's me," said Bouwmeester, pointing to a youthful version of himself. And pointing to Penrose, he said, "Roger doesn't change, but I do." Indeed, Penrose looked no different from when I met him more than fifteen years after the picture was taken.

The problem they needed to solve, to do the experiment on an optical bench on Earth, was essentially the one that Penrose solved by going up into space: how to keep the photon in a superposition for a sufficiently long time to witness any potential collapse of the movable mirror's superposition. On Earth, they needed to store a photon awhile before letting it come back into the interferometer and head toward the beam splitter. One option was to store the photon in an optical cavity, which is essentially made of two extremely high quality concave mirrors aligned such that the photon, once it enters the cavity, keeps bouncing back and forth between the mirrors and then, at some random time, leaks out again. It's a way of storing the photon for a certain length of time.

So the photon's journey to satellites B and C is replaced by two optical cavities, each holding on to the photon for a while, as if it were traveling 10,000 miles and back. The optical cavity that replaces the part of the photon path with the movable mirror is somewhat unique. In this cavity, one of the mirrors is tiny and suspended on a cantilever arm. Bouwmeester decided to use optical and infrared photons—it's easier to make high-quality mirrors for them than it is for X-ray

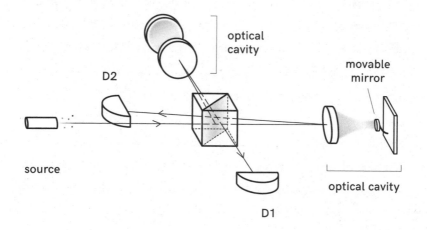

optical cavity

D2

movable mirror

source

optical cavity

D1

photons, which Penrose had used in his original thought experiment. The photons bounce between the two mirrors inside the optical cavity. This creates a "radiation pressure" that is strong enough to displace the movable mirror. This phenomenon is itself quite curious. Quantum mechanically speaking, the photon is not localized. It's all over the place inside the cavity, and over time, its delocalized presence creates the necessary pressure to push at the mirror.

Now, just as the photon is in a superposition of being in two arms of the interferometer, the movable mirror ends up in a superposition of being displaced and not-displaced.

At some random moment, the photon leaks out of the cavity and heads back toward the beam splitter. What happens next depends on whether the photon, and indeed the entire system (including the movable mirror), is still in a coherent state of superposition or has collapsed to one or the other state.

If the entire system is still in superposition, the photon's two states will interfere. Since one of the path lengths is fixed, the interference pattern will depend on the position of the movable mirror at the exact time the photon leaks out—it dictates the distance traveled by the photon in the arm of the interferometer with the movable mirror. The interference pattern created over many runs of this experiment— detected as the number of clicks at detectors D1 and D2—will have a signature that's tied to the movable mirror's oscillations.

But if the mirror's superposition has collapsed, the photon will act like a particle and has an equal chance of heading toward D1 or D2. As with Penrose's space-based experiment, monitoring the statistics of the detections at D1 and D2 can tell us whether the tiny mirror has remained in superposition or not.

The fundamental task at hand in such an experiment is to put a macroscopic object into superposition and keep it there long enough to do an experiment. When they wrote their paper in 2002, Bouwmeester and Penrose claimed that a tiny mirror could be put into a superposition of being at two locations if one could combine the state-of-the-art technologies for each piece of the puzzle. "That's still true, but it is extremely hard to combine state-of-the-art low temperature physics with state-of-the-art optics with state-of-the-art mechanical fabrication and so forth," Bouwmeester said. "That's basically what we have been working on ever since."

It turns out that none of these issues is trivial. Far from it. First they had to learn how to fabricate mirrors several orders of magnitude smaller than a grain of sand. One technique involved using a focused beam of ions to cut out a mirror and then glue it to a tip of a cantilever, to make it movable; the mirror was so small and so hard to control that oftentimes it'd flip during the fabrication process and get glued upside down. Even if it was right side up, such a mirror was still too large. The team figured out how to make smaller mirrors cantilevered at the tip of slivers of silicon nitride. They also had to make these mirrors unbelievably cold. Otherwise the thermal jiggling of the molecules of the mirror would be such that the impact of a single photon would have no discernible effect. So the mirrors had to be cooled down to bring them to their quantum ground state—which meant achieving temperatures below 1 millikelvin. "That's ridiculously low for a cryogenics experiment," said Bouwmeester. But getting things so cold means using dilution refrigerators, pumps circulating helium, and the like—all sources of vibration that could render the whole exercise futile. And, therefore, they had to develop multiple systems to dampen vibrations. Of course, all of this has to be done inside a vacuum chamber. "In the end this apparatus costs several millions," said Bouwmeester. All for a tiny mirror that stays so cold and quiet that it can be pushed around by a photon and end up in a superposition of two positions, one position that's barely a few hundred atomic widths away from the other.

"You first have to prove that you can create a quantum superposition of a macroscopic object before you can investigate its decoherence," said Bouwmeester. "In the end, we are still quite far away from it. [But] the developments have been enormous."

Decoherence specifically refers to the loss of coherent superposition of a quantum mechanical system due to its interaction with the environment, such that it ends up in some classical state. Penrose's ideas and GRW-like collapse theories are not theories about decoherence: they explicitly advocate collapse, which leads to decoherence.

As we talked, Bouwmeester admitted that despite being inspired by Penrose's ideas to carry out such difficult experiments, he thinks the experiments will most likely *not* see any collapse of quantum superpositions of more and more massive objects, as long as the objects are well-enough isolated from their environments to prevent decoherence. In which case, Bouwmeester says that he'd be forced to take seriously the idea that there is no quantum-classical boundary, that wavefunctions evolve and there is no collapse. Different parts of the wavefunction continue evolving, and as they interact with their environment, they behave as if decoherence has set in, making it difficult if not impossible to get the separately evolving wavefunctions to interact. "They become independent and there is no interference anymore," said Bouwmeester. "But this is rather strange, because then you are really back to Schrödinger's cat."

Yes, back to the poor cat, but in a subtly different way. Instead of being a demonstration of the absurdity of quantum mechanics, it becomes an unflinching exploration of its implications. In this way of thinking, Bouwmeester is hinting at the argument that both the dead cat and the live cat exist, and so does someone who has seen a dead cat and someone who has seen a live cat. They are two distinct minds, and there are possibly two different worlds that these two minds inhabit, which don't interact anymore. "That's not a ridiculous way of looking at things," he said. "You just have to go through quantum mechanics for a little while to understand how elegant and simple it actually is."

Some physicists take the simplicity and elegance of quantum mechanics to heart, such as the straightforward evolution of the wavefunction according to Schrödinger's equation and the attendant superpositions, and refuse to add anything to its formalism, even the notion of collapse due to measurement, which is a modification of the Schrödinger evolution. And they end up with a startling conclusion: superpositions of systems that cannot interfere with each other anymore now each exist in their own right. The idea leads us to a notion of "many worlds," where every possibility exists somewhere. For Bouwmeester, if experiments like his never see collapse, even as the macroscopic objects in superposition keep getting bigger and bigger, that's a sign. "In that case I am really going to take the many worlds interpretation seriously," he said.

Bouwmeester first realized just how earnestly some physicists regard the many worlds interpretation when he met Lev Vaidman (of Elitzur-Vaidman bomb puzzle fame) on a bus in China, on their way to a conference. Vaidman has famously written in one of his papers, "The collapse [of the wavefunction] . . . is such an ugly scar on quantum theory, that I, along with many others, am ready to . . . deny its existence. The price is the many-worlds interpretation (MWI), i.e., the existence of numerous parallel worlds."

"He was rather upset when I met him," Bouwmeester said, speaking at the Institute for Quantum Computing in Waterloo, Canada. Vaidman, it seems, had been trying to get a patent approved for a watch that would help him make a difficult "yes or no" life decision. The watch would have a single photon source. The photon would go through a beam splitter and be detected by one of two single-photon detectors inside the watch. If one of them clicks, the watch says "YES," do it; if the other clicks, the watch says "NO," don't. Vaidman's point being that no matter what decision you make, you can rest easy because you know that in another branch of the wavefunction, you have done the opposite.

One person who likely would have been unfazed by Vaidman's watch is Hugh Everett III, a mathematician and quantum theorist who first advocated, in his PhD thesis in 1957, taking the collapse-free evolution of the wavefunction seriously, as a way of solving the measurement problem. His thesis led to possibly the most unsettling solution to the paradox of the double-slit experiment yet entertained—the many worlds interpretation.

8

HEALING AN UGLY SCAR

The Many Worlds Medicine

Actualities seem to float in a wider sea of possibilities from out of which they were chosen; and *somewhere*, indeterminism says, such possibilities exist, and form a part of the truth.

—William James

If there is one place that could be said to have harbored a handful of quantum dissenters—those who found the Copenhagen interpretation problematic, if not distasteful—it'd have to be Princeton, New Jersey. Einstein, the original dissenter, came to the Institute for Advanced Study in 1933 and lived out the rest of his life there, and remained forever of the opinion that quantum mechanics wasn't complete. David Bohm, who came to Princeton University in 1946, started thinking contrarian views there, before going into exile in Brazil in 1951, from where he published his hidden variable theory. Soon after Bohm left Princeton, a mathematically minded young man named Hugh Everett III, having just gotten a bachelor's degree in chemical engineering, came to Princeton University in 1953 and by 1955 had begun working on his PhD in quantum physics. His supervisor was John Wheeler. Though Wheeler was a staunch supporter of Niels Bohr and the Copenhagen interpretation, Wheeler's protégé would turn out to be one of the most imaginative of the nonconformists.

Wheeler put a lot of stock in taking the equations of physics seriously and seeing where they led us. Soon after Einstein came up with his general theory of relativity, solutions of his equations were pointing physicists toward topological structures in spacetime that taxed common sense. In the 1960s, Wheeler would coin the terms *black hole* and *wormhole* for such structures. But even earlier, Wheeler's attitude likely rubbed off on Everett—and he'd apply it to the mathematics of quantum physics.

It started with taking seriously the wavefunction and its evolution, in all its simplicity and elegance. The essence of Everett's thinking was that the wavefunction is all there is: a *universal wavefunction* for the entire universe, which describes the universe as being in a superposition of any number of classical states, and this wavefunction and the superpositions evolve continuously, deterministically, and forever.

Everett's intuition was informed by the need to do away with the measurement problem. By 1955, he had identified what he thought of as the key issue with the quantum formalism then in vogue. If the state of a quantum system at any instant is given by the wavefunction psi, ψ, then Everett pointed out that there are two processes that govern it. First, the wavefunction evolves in time according to

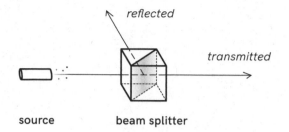

reflected

transmitted

source   beam splitter

Schrödinger's equation, a completely deterministic process. But upon measurement, the wavefunction abruptly changes to a definite state with some probability that can be calculated—a so-called probabilistic jump. Everett found this untenable.

He asked if these two processes were compatible with each other. More specifically, he asked, "What actually does happen in the process of measurement?"

Consider a photon that goes through a beam splitter. According to standard quantum mechanics, it goes into a superposition of being in two paths. The wavefunction of the photon at this point is a linear combination of two wavefunctions, one in which the photon takes the reflected path, and the other in which it takes the transmitted path. (As we saw earlier, if $\psi = a.\psi_{ref} + b.\psi_{tr,}$ where the coefficients "a" and "b" are complex numbers, then the square of the modulus of a, or $|a|^2$, is the probability of finding the photon in the reflected path, and $|b|^2$ the probability of finding it in the transmitted path. If the beam splitter is built to reflect half the light and transmit half the light, then the probabilities are each equal to 0.5.)

Now, if you have detectors D1 and D2, one at the end of each path, then for each photon that goes through the beam splitter, we'll get a click at either D1 or D2. In the Copenhagen interpretation,

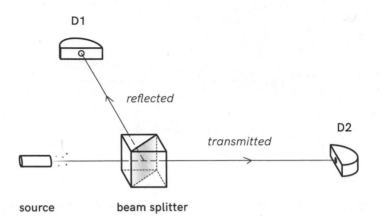

because the detectors are somehow magically treated as classical objects, the measurement causes the collapse of the wavefunction, and the photon is localized at either D1 or D2. Recall that in the 1960s, Eugene Wigner too found this distinction between the quantum and the classical rather arbitrary, and argued that it's the consciousness of an observer that causes the collapse.

Everett took a different tack. If you follow the math and treat the detectors quantum mechanically too, then the entire apparatus ends up in a superposition of D1 clicking *and* D2 clicking. Why just the detector? Why not also treat the observer quantum mechanically? If so, the observer ends up in a superposition of hearing D1 click *and* hearing D2 click. According to Everett, if you consider only a part of this wavefunction in which D1 clicks, you are left with a *definite observer* who hears the click. And examining another part of the wavefunction reveals a definite observer who hears D2 click. "In other words, the observer himself has split into a number of observers, each of which sees a definite result of the measurement," he wrote.

What Everett was proposing is that there is no collapse of the wavefunction and hence no measurement problem. All possibilities exist (and we'll soon come to what *exist* might mean, again). In the simple case of a beam splitter and two detectors, we end up with two observers, each of whom hears one of the detectors click.

Now think of an observer in one of those branches. Having observed, say, D1 click, the observer sends another photon through the same beam splitter. Again, the observer splits, into one who hears D1 click and one who hears D2 click. There's no collapse, no probabilistic jump, just the continuing evolution of the wavefunction. This process can be carried on ad infinitum—and we get a treelike structure of observers. If you follow any one branch of this tree, you will find an observer who hears the detectors click, for example, in the following sequence: D1, D1, D2, D1, D2, D1, D1, D2 . . . ; or D1, D1, D1, D2, D1, D2, D1, D1 . . . ; or a seemingly random combination of D1s and D2s.

So, even though there is no probabilistic jump occurring in any one such sequence of detections—in the sense of the wavefunction collapsing randomly to one or other state—for each observer *it seems* as if D1 or D2 clicks randomly, and thus there is a perception of a collapse to one state or the other. Everett argued that "for almost all of the 'branches' of his 'life tree,'" an observer would hear D1 or D2 clicking at a frequency that would tally with the probabilities given by the initial superposition, provided of course one carried out this experiment enough times (it'd turn out to be not quite so simple).

Everett was proposing a theory that was continuous (there were no real jumps, just apparent jumps) and causal, because everything evolved deterministically according to the rules of the Schrödinger equation. Yet, for any given observer, the theory is discontinuous because of the perceived jumps in states, and the jumps are seemingly random. The theory, Everett wrote, "can lay claim to a certain completeness, since it applies to all systems, of whatever size . . . The price, however, is the abandonment of the concept of the uniqueness of the observer, with its somewhat disconcerting philosophical implications."

He even came up with an analogy to drive home the point: "One can imagine an intelligent amoeba with a good memory. As time progresses the amoeba is constantly splitting, each time the resulting amoebas having the same memories as the parent. Our amoeba hence does not have a life line, but a life tree. The question of the identity or non identity of two amoebas at a later time is somewhat vague. At any time we can consider two of them, and they will possess common memories up to a point (common parent) after which they will diverge according to their separate lives thereafter . . . The same is true if one accepts the hypothesis of the universal wavefunction. Each time an individual splits he is unaware of it, and any single individual is at all times unaware of his 'other selves' with which he has no interaction from the time of splitting."

Wheeler was impressed by Everett's work, and yet he had serious reservations about these "disconcerting philosophical implications." Everett was taking on the Copenhagen interpretation, and as such, Wheeler—who admired Niels Bohr—wanted to discuss Everett's work with those in Copenhagen. But Wheeler had concerns about splitting observers and amoebas. "I am frankly bashful about showing it to Bohr in its present form, valuable & important as I consider it to be, because of parts subject to mystical misinterpretations by too many unskilled readers," Wheeler told Everett.

Everett avoided some of these "mystical" overtones in his thesis, especially the bit about splitting amoebas, and then submitted it to Wheeler in 1956. While expounding his ideas in great detail and with mathematical rigor, he nonetheless took aim at Bohr and the Copenhagen interpretation, calling it conservative and overcautious (quite the irony, when you consider that the Copenhagen view, in its extreme, says that reality doesn't exist until one observes it). "We do not believe that the primary purpose of theoretical physics is to construct 'safe' theories at severe cost in the applicability of their concepts, which is a sterile occupation, but to make useful models which serve for a time and are replaced as they are outworn," wrote Everett. He criticized the Copenhagen interpretation for relying on a form of "objectionable" dualism, splitting the world into the classical and the quantum, and ascribing to the classical world a reality that it denied to the quantum.

As expected, Everett's thesis was received rather coldly in Copenhagen. An American physicist, Alexander Stern, who was in Copenhagen at the time, organized a seminar in May 1956 in which he, Niels Bohr, and others discussed Everett's ideas. Stern wrote a letter to Wheeler a week later, in which he detailed the criticisms of the Copenhagen crowd, taking issue particularly with Everett's idea of a universal wavefunction. Stern said that Everett's ideas "lack meaningful content" and said that some aspects were a "matter of theology."

Wheeler wrote back almost immediately, and was even somewhat apologetic. "I would not have imposed upon my friends the burden of analyzing Everett's ideas, nor given so much time to past discussions of these ideas myself, if I did not feel that the concept of 'universal wavefunction' offers an illuminating and satisfactory way to present the content of quantum theory." He then both praised Everett and mischaracterized his position: "... this very fine and able and independently thinking young man has gradually come to accept the present [Copenhagen] approach to the measurement problem as correct and self consistent, despite a few traces that remain in the present thesis, *draft of a past dubious attitude* [italics mine]."

Everett did no such thing. He did rework his thesis into a version that was almost three-quarters shorter (Wheeler had asked him to make it "javelin proof"); he took out the sharpest attacks against the Copenhagen interpretation, including his denunciation of the measurement problem, and recast his views about quantum mechanics as a way to solve the problem of reconciling general relativity with quantum mechanics into a theory of quantum gravity. But the underlying mathematical formalisms in the long and short versions of the thesis were essentially the same. His views on the Copenhagen interpretation had not changed, something that became abundantly clear in a correspondence with the theoretical physicist Bryce DeWitt.

DeWitt had edited the issue of *Reviews of Modern Physics* in which Everett's shortened thesis had appeared. DeWitt would later say about Everett's paper: "I was stunned, I was shocked." DeWitt wrote to Wheeler, raising some concerns, including the issue of the splitting of observers: "I can testify to this from personal introspection, as can you. I simply do *not* branch." Wheeler forwarded the letter to Everett.

In his reply, Everett called the Copenhagen interpretation

"hopelessly incomplete" and "a philosophic monstrosity with a 'reality' concept for the macroscopic world and denial of the same for the microcosm."

Everett also clearly outlined what happens to the various macroscopic superpositions in the universal wavefunction: "From the viewpoint of the theory, all elements of a superposition (all 'branches') are 'actual,' none any more 'real' than another. It is completely unnecessary to suppose that after an observation somehow one element of the final superposition is selected to be awarded with a mysterious quality called 'reality' and the others condemned to oblivion. We can be more charitable and allow the others to coexist—they won't cause any trouble anyway because all the separate elements of the superposition ('branches') individually obey the wave equation with complete indifference to the presence or absence ('actuality' or not) of any other elements." In other words, to answer DeWitt, one version of "you" does not interact with any other version of "you" after splitting, so you can never feel yourself splitting.

But a more seemingly preposterous idea was waiting in the wings to be unleashed: that somehow each splitting is causing the universe itself to divide into parallel worlds. Everett would bring up the idea at a conference in October 1962 in Cincinnati, Ohio. Among the participants were many luminaries, including Nathan Rosen, Boris Podolsky, Paul Dirac, Abner Shimony, and Eugene Wigner. Everett was there too. The attendees began raising the uncomfortable question of parallel universes. At one point, Shimony said that the idea that all macroscopic superpositions continue to exist, even for a single observer, had strange consequences. "It seems to me that if this is the case, there are two possibilities. The two possibilities

involve awareness. One possibility is that ordinary human awareness is associated with one of these branches and not with the others. Then the question becomes, how does your formalism permit
this solution? The other possibility is that awareness is associated
with each branch."

Podolsky said, "Somehow or other we have here the parallel
times or parallel worlds that science fiction likes to talk about so
much. Every time a decision is made, the observer proceeds along
one particular time while the other possibilities still exist and
have physical reality." To which Everett replied, "Yes, it's a consequence of the superposition principle that each separate element of
the superposition will obey the same laws independent of the presence or absence of one another. Hence, why insist on having a certain selection of one of the elements as being real and all of the
others somehow mysteriously vanishing?"

After some back-and-forth, Shimony said to Everett, "You eliminate one of the two alternatives I had in mind. You *do* associate
awareness with each one of these." Everett concurred: "Each individual branch looks like a perfectly respectable world where definite
things have happened," he said. This was about the only instance in
which Everett explicitly acknowledged the notion of multiple worlds.

It was DeWitt who eventually gave credence to the idea of many
worlds. In an article for *Physics Today* in 1970, DeWitt explained the
new interpretation, in which a universe is represented by a single
wavefunction. "This universe is constantly splitting into a stupen-
dous number of branches, all resulting from the measurementlike
interactions between its myriads of components. Moreover, every
quantum transition taking place on every star, in every galaxy, in

every remote corner of the universe is splitting our local world on earth into myriads of copies of itself."

DeWitt would later remember his own astonishment at the realization: "I still recall vividly the shock I experienced on first encountering this multiworld concept. The idea of 10^{100+} slightly imperfect copies of oneself all constantly splitting into further copies, which ultimately become unrecognizable, is not easy to reconcile with common sense. Here is schizophrenia with a vengeance."

Despite his shock and awe, DeWitt was a convert and he'd play a major role in proselytizing Everett's interpretation, which goes by many names now, but we'll refer to it as Everett's many worlds or simply the many worlds interpretation.

The first thing that greets you as the elevator doors open on the fourth floor of the Downs-Lauritsen Laboratory of Physics at Caltech is a giant mural of Feynman diagrams—the kind of squiggly drawings that Feynman would draw on paper napkins to visualize the interaction of particles. I was there to meet theoretical physicist Sean Carroll, a proponent of the many worlds interpretation. We were midway through our discussion about quantum mechanics when Carroll decided he was going to split the world.

His iPhone has an app called the Universe Splitter, which is a version of the watch that Vaidman wanted to patent—one that will help you make up your mind when confronted with a difficult YES or NO decision. There is no wrong decision, for—in the Everettian view—there exists a universe in which the app suggests a different decision. So why worry?

Carroll fired up the app, with its default choices of what to do:

Take a chance or *Play it safe* (we could have typed in something else, but we stuck with those choices). Carroll pressed a button that said, ominously, "Split Universe." The app sent a command to a lab somewhere near Geneva, Switzerland, where a single photon was sent through a beam splitter. "If you believe in Everett, there is a world in which the photon goes left and a world in which the photon goes right," said Carroll.

A few seconds later, the result came back. "Ah, we are in the universe where we have to take a chance." And the act of saying aloud the words *Take a chance* (and presumably the words *Play it safe* in another world) had split the universe irreconcilably (we'll come to why in a moment). "Now there are just two copies of me."

And presumably me, I thought. Was this real or surreal?

What I had witnessed was an experimental realization of one part of the Mach-Zehnder interferometer, which becomes a strangely simple thing to behold in the many worlds interpretation.

Consider first a configuration with just the first beam splitter, and nothing else. The Everettian view is that the wavefunction of the universe now has two components: one in which the photon is transmitted, and one in which the photon is reflected (let's assume for now that this is the only quantum choice being exercised in the

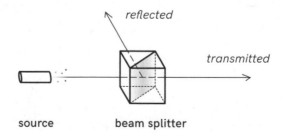

entire universe). So far, both the Copenhagen interpretation and the many worlds interpretation are in sync. The system is now in a superposition of those states and continues to evolve according to the Schrödinger equation.

If we now put detectors at the ends of each path, the two interpretations give us wildly different views of reality.

In the Copenhagen view of things, one of the detectors clicks. If the measurement apparatus is considered classical, then that click, so to say, is the sound of the wavefunction collapsing.

For a while, physicists thought they had come up with a way to explain this collapse without resorting to magic. Let's say the detector can also be treated as a quantum mechanical object. If it's not kept completely isolated, the detector eventually starts interacting with its environment, mainly through other ambient photons and air molecules bouncing off it, and the detector becomes entangled with the environment. It's mathematically impossible to describe this complicated interaction. So, what quantum mechanics does is to describe the combined state of the photon and the detector using

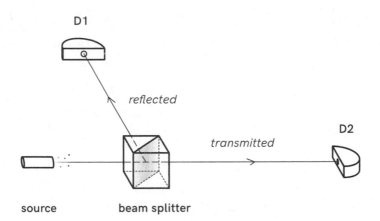

something called a density matrix—basically, a mathematical formalism that ignores the environment. Before the interaction with the environment, the photon and the detectors are in a definite state of "the photon is reflected and D1 clicks" *and* "the photon is transmitted and D2 clicks." After the interaction with the environment, the quantum formalism says that the system is in a state of either "the photon was reflected and D1 clicked" *or* "the photon was transmitted and D2 clicked," but we don't know which. The latter situation now represents a state that encodes our ignorance of what happened.

"If you ignore the environment, the best you can say about your quantum system and the measuring apparatus is that they are in a mixed state, described by a density matrix," said Carroll.

The density matrix allows you to calculate the probability that D1 clicked or D2 clicked (0.5 each, for this experiment). In this case, the probabilities look like classical probabilities, in that they are grounded in our ignorance. The process of interaction with the environment is called decoherence, and the fact that the resultant density matrix lets you calculate the correct probabilities led physicists to think that decoherence—when it was first proposed—actually caused the collapse of the wavefunction and thus solved the measurement problem. But that excitement was short-lived. Decoherence, while it says that the combination of a quantum system and the measuring apparatus evolves to look like a system in a probabilistic mixture of classical states, doesn't really explain why.

In the classical world, when we use probabilities to talk about the state of a system, it's because we are ignorant, but the system nonetheless is in some definite state, and there is nothing else it's interacting with. In the quantum world, the probabilities calculated using the density matrix are somewhat different. It appears as if we are ignorant of the exact state, but unlike the classical state, the quantum state being described by the density matrix is not a definite state. We'd have to consider the entanglement with the outside world to describe the quantum state in its entirety, and the density matrix doesn't do that.

So the theory of decoherence comes tantalizingly close to making sense of collapse, but fails. The Copenhagen interpretation drops the ball at this point, whereas the many worlds interpretation picks it up and runs with it.

According to many worlds, both D1 and D2 click, each in their own branch of the wavefunction. This clicking and the consequent interactions with their local environments lead to entanglement and decoherence. Once decoherence sets in, the two worlds begin evolving independently, but still according to the Schrödinger equation. However, the two branches of the evolving wavefunction are now impossible to recombine. "The environments attached to each decoherent branch are orthogonal to each other, which means that there will never be any interference," said Carroll. So, from the point of view of any one branch of the wavefunction, "all the other branches are still there, they are just exponentially hard to find."

But let's say that you did not put detectors D1 and D2 at the end of each path coming out of the beam splitter, thus avoiding decoherence. Then, in principle, it's possible to recombine the two worlds by bringing the two paths together at a second beam splitter. That is exactly what happens in a Mach-Zehnder interferometer. In order to calculate the probability of finding the photon at D1 or D2, after it crosses the second beam splitter, we have to take into account that it went through both paths, a different world for each path.

This is reminiscent of Feynman's approach to solving the puzzle of the double-slit experiment, or quantum mechanics more generally. Feynman came up with what he called the path integral formulation of quantum mechanics. In this approach, a particle approaching a double slit can still be treated classically—in that it goes through one or the other slit, but in order to calculate the probability that it lands on a particular place on the far screen, you have to let the particle take *every possible* path from the two slits to the screen. These paths include all sorts of squiggly trajectories that don't make any classical sense. Each of these paths is assigned a weight that dictates its contribution to the final probability.

"What quantum mechanics tells us fundamentally about how to think about the universe is that in order to calculate the probability of something happening we have to add the amplitudes for all the different ways it could occur," Aephraim Steinberg told me. When you do that, you get interference. "The insight that Feynman had was to realize that what's really interfering are two different states of the universe. And in the simplest case, those two states might only differ by where a single particle is. Is the electron in the upper path or the lower path?"

While Feynman's path integral approach is a tool for calculating the probabilities of experimental outcomes in this world, the many worlds approach takes this idea of different states of the universe rather more literally. And this, according to Carroll, is what makes its take on reality very appealing, and understanding the double slit a breeze, assuming, of course, that you are not perturbed by the idea of new branches of the wavefunction and hence new worlds appearing at every quantum fork in the road. "There is a heavy psychological price to pay, and the question is, how much does that bother you?" said Carroll. "Doesn't bother me at all."

He is far more bothered by the Copenhagen interpretation. Take, for instance, Bohr's language that's used to describe the double-slit experiment: if we don't collect which-way information, the photon behaves like a wave; if we do, it behaves like a particle. "All that is complete nonsense," said Carroll.

As an Everettian, Carroll thinks simply in terms of a wavefunction that splits into two components at a beam splitter, and each continues to evolve according to the Schrödinger equation. If there's no decoherence, then those two parts of the wavefunction can be made to interfere. "If you don't untangle the photon as it moves through the slits, then you'll see the interference pattern, because that's the solution of the Schrödinger equation," said Carroll.

There's no doublespeak, so to say, about the photon showing its wave nature or particle nature. It's just the wavefunction evolving and doing its thing—and the wavefunction represents the quantum state of reality. "Many abstract thinking physicists are . . . impressed by the underlying mathematical beauty and elegance of Everett," said Carroll. "Physicists are suckers for mathematical beauty and elegance."

There's certainly a mathematical simplicity to the many worlds idea. There's just the wavefunction and its evolution. No added ingredients (such as hidden variables) or ungainly nonlinear dynamics (such as stochastic collapse à la Penrose or GRW) or Copenhagen-like magic to induce collapse.

This point was brought home vividly when I visited philosopher David Wallace, whose office is about fifteen miles away from Carroll's, at the University of Southern California, Los Angeles. Wallace, who had previously worked with David Deutsch at the University of Oxford, had just moved to California. And like Deutsch, Wallace is a strong proponent of the many worlds interpretation. He began his academic career as a theoretical physicist but then switched to philosophy (when theoretical physics "started to sound a little bit too practical," he's known to joke). And as a philosopher, he was seduced by Everett's many worlds interpretation.

"To me, one of the most attractive things about the Everett interpretation is that it doesn't commit you to a revisionary project in physics," Wallace told me. "I'm very skeptical that a revisionary project would succeed."

To make his case, Wallace tore a page off his yellow, lined writing pad, and sketched a 2-D coordinate system, with the X-axis representing "Change the physics?" and the Y-axis representing "Change the philosophy?". The positive part of each axis represented YES and the negative part represented NO.

The physics here refers to the evolution of a physical system according to the standard Schrödinger equation. The philosophy refers to the way we do science: standard scientific realism, the idea that our theories are an objective, observer-independent description of the reality that is out there.

He crosshatched the quadrant that involved YES for both.

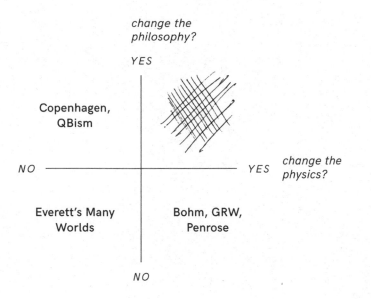

*change the
philosophy?*

YES

Copenhagen,
QBism

NO ——————————————+—————————— YES   *change the
physics?*

Everett's Many   Bohm, GRW,
Worlds   Penrose

NO

"Nobody would want to do both of them, so cross out that box," he said.

Copenhagen and Quantum Baycsianism (QBism) went into the NO, YES quadrant: they don't change the physics, but they change the philosophy, because the interpretations are not observer-independent (however you define an observer). Copenhagen does involve a collapse, which is non-Schrödinger evolution, but since it does claim a law for how that happens, one can argue that it does not modify the physics.

Bohmian mechanics, GRW, and Penrose's collapse theory all modify the physics, either by adding hidden variables or by adding new dynamics that interrupt the Schrödinger evolution of a system, causing it to collapse. But they leave the philosophy alone.

Everett modifies neither the physics nor the philosophy. "This sounds weird for something as crazy as the Everett interpretation, but the attraction for me is that it's extremely conservative," said Wallace.

If the many worlds interpretation is such a thing of beauty and elegance, then how come not everyone is sold on the idea? For starters, there's the obvious discomfort with the thought of, well, many worlds. This was voiced early in the life of the Everett interpretation, most notably by Abner Shimony at the October 1962 meeting in Cincinnati, Ohio. Shimony said, "I think one should invoke Occam's razor: Occam said that entities ought not to be multiplied beyond necessity. And my feeling is that among the entities which aren't to be multiplied unnecessarily are histories of the universe. One history is quite enough." Change the word *histories* to *worlds*, and the objection gets even more trenchant.

Of course, the proponents think that the naysayers have applied Occam's razor to the wrong issue. The physicist Paul Davies once asked David Deutsch, "So the parallel universes are cheap on assumptions but expensive on universes?" Deutsch said, "Exactly right. In physics we always try to make things cheap on assumptions."

Wallace is of the same opinion. Yes, the Everettian interpretation argues for the existence of many histories, worlds, universes (however you want to think of what's happening). "But the crucial thing is that you didn't add that material to the underlying equation. You just interpreted the equations that way," said Wallace. "I don't think there is a particularly defensible scientific principle that says that fewer things are better. There is a very defensible scientific principle that says simpler things are better. The many worlds interpretation mathematically is uncontentiously simpler than any of [the other] modifications."

And from a cosmological point of view, the idea that the many worlds interpretation requires too much stuff is also batted away by Carroll and Wallace. Carroll, who is a cosmologist, likes thinking in terms of a wavefunction for the entire universe and its evolution independent of observers—it allows him to deal with the physics of the big bang and black holes, for example. Also, cosmologists think that our unobserved universe is in any case far, far larger than what we can see through our telescopes. "We are already, in physics, committed to incredibly large amounts of stuff," said Wallace. What's some more of the same?

There are other arguments in defense of many worlds. In the mathematical formalism of quantum mechanics, a quantum state can be represented by a vector in a coordinate system called the Hilbert space. A vector in 2-D space is a directed arrow that begins at the origin (0,0) and goes to a point (X,Y). Similarly, a vector in 3-D space begins at the origin and ends at some point (X,Y,Z). A vector in Hilbert space is conceptually the same, except it's dimensionality can be enormous. And in the Everettian way (or indeed when you contemplate the state of the entire universe in any interpretation), the wavefunction of the universe is a vector in the Hilbert space for the entire universe, and the dimensionality of this abstract mathematical space is mind-bogglingly large. "It could be infinite, but even in the most pessimistic readings, it is something like e to the power of 10^{120}, which is just an enormously crazy number," said Carroll (where e equals 2.72, approximately). "There's plenty of room for many, many branchings. We are nowhere near done."

Schrödinger evolution merely tells you how the quantum state of the universe, represented by one vector in this Hilbert space, changes to another vector. So if a wavefunction denoted by one vector splits into two vectors, does each vector represent a physical universe? "People fret about that, but I think it's fine. I have no trouble thinking of them as universes," said Carroll. "They are not located in our physical space, they are separate copies of our physical space, located in Hilbert space."

Similar arguments are also used against those who say that the many worlds interpretation flouts laws of conservation of energy. Where does the energy for the new physical branches come from? Well, all these worlds/universes exist in Hilbert space—not in physical space—so the question is a bit ill posed. Nobel laureate Frank Wilczek has argued, for instance, that "if the other universes are inaccessible, they cannot be sources or sinks of energy."

To Carroll, worrying about the number of worlds is fruitless. "Let's grow up and move beyond that," he said.

That's because there are other seemingly more pressing and legitimate concerns about the Everettian view. One is about trying to figure out what exactly happens when a universe splits. Say we send a photon through a beam splitter and let each path decohere, resulting in two separate worlds. Does the entire universe split into two everywhere at the same instant (and what does that mean, given that Einstein's relativity abolished the notion of a universal "now") or does it start splitting at the point where the decoherence happens near the beam splitter, and move outward at the speed of light? Opinions differ, and there's no consensus, even among those who are not troubled by the idea of many worlds.

Perhaps the most well known concern about the Everettian view has to do with the meaning of probability, which is contested in physics and science in general. But it comes to the fore in the many worlds interpretation. "It just brings into the open the inherently mysterious nature of probability," said Wallace.

Let's say you set up a Mach-Zehnder interferometer with path lengths such that if you sent photons one by one into the interferometer, 75 percent of them will end up at detector D1 and 25 percent at detector D2 (recall that the difference in path lengths can be tuned to get this result). The wavefunction of the photon after it has crossed the two beam splitters can be written as a linear combination of two wavefunctions ($\psi = a.\psi_{D1} + b.\psi_{D2}$, where a and b are amplitudes, and $|a|^2$ equals 0.75 and $|b|^2$ equals 0.25, and these numbers are to be taken as the probabilities of detecting the photon at D1 and D2, respectively, the so called Born rule). "The question is why do amplitudes-squared get interpreted as probabilities of anything?" said Carroll.

In the Copenhagen interpretation, that's just by decree: randomness is inherent in reality and the Born rule gives us the probabilities of measurement outcomes. In Bohmian mechanics, even though the entire evolution of the quantum system is deterministic, the probabilities arise because we are uncertain of the initial conditions. In collapse theories, there is irreducible randomness in the dynamics of a quantum system and there is really something stochastic happening at the microscale, regardless of measurement.

In the many worlds interpretation, the waters get muddied somewhat. In Everett's original view, each time you send the photon into our 75-25-tuned interferometer, the universe splits into two:

one in which D1 clicks and D2 doesn't, and another in which D2 clicks and D1 doesn't. So in one world D1 clicks with a probability of 1, and in the other world D2 clicks with a probability of 1. Significantly, both worlds are real. Then what's one to make of the probabilities of 0.75 and 0.25 assigned to these outcomes by quantum mechanics? One tack is to imagine that the experiment continues in each new branch of the universe, and splitting continues too, creating more branches. After a large, potentially infinite number of observations, one can look at the frequency of clicks of D1 and D2 in each branch and ask: do the frequencies come close to the ideal of 75 percent for D1 and 25 percent for D2? Not quite. "It doesn't work if you just count up the worlds," says the Australian physicist Howard Wiseman, whose work we encountered in the context of weak measurements and Bohmian trajectories. "In the vast majority of worlds, the relative frequencies are nothing like the quantum probabilities." Everett used a sleight of hand to argue that some of these worlds should be disregarded—an argument that relies on the Born rule. In the remaining worlds, given an extremely large number of observations, probabilities can be thought of as frequency of outcomes. "But then what happened to the idea that all worlds are equally real?" says Wiseman. "How come you are now effectively throwing away almost all of them, as if they are not as good as the others?"

He's not the only one troubled by this way of linking frequency of outcomes to probability in the many worlds scenario. Carroll and Wallace are too.

Carroll suggests one way out: think of probability as something subjective. Carroll and philosopher Charles "Chip" Sebens have argued that $|a|^2$ and $|b|^2$ should be interpreted as numbers that represent our uncertainty about the outcome of a measurement. So, as in classical physics, probabilities here are due to our ignorance, except in this case the ignorance is about something quite dramatic: we don't know which branch of the wavefunction we are on. Let's say you did one run of the experiment, sending one photon through the interferometer. D1 clicks in one branch and D2 clicks in the other branch of the wavefunction, decoherence ensues, and you'll soon enough find yourself in a branch of the universe in which either D1 clicked or D2 clicked. "The branching happens first, because decoherence is very, very fast, on microscopic time scales, [of] 10^{-20} seconds or less. There is always a period of time in which the branching has occurred and there are two copies of you, but those two copies are exactly the same, because they don't know what branch they are on yet," said Carroll. Thus, even though there are two copies of you, for a very tiny time period, those two copies are ignorant about the branching, and it's this ignorance that explains the outcomes of experiments in terms of probabilities. Carroll and Sebens have shown that in that brief moment, post-decoherence, if you were to assign probabilities to D1 clicking or D2 clicking, under certain simple assumptions, you'd end up with $|a|^2$ and $|b|^2$, respectively: which is the Born rule. "There's a real world," but we are uncertain about where we are in that real world, said Carroll.

There's yet another way to think of probabilities in many worlds. Wallace uses decision theory, an approach pioneered by David Deutsch, which is the study of the reasons behind the choices one makes or the bets one places. If you were doing the above experiment and had to bet on the outcome of the measurement, then, according to Wallace, the rational thing for you to do before the experiment is to treat $|a|^2$ and $|b|^2$ as probabilities to place your bets on which branch of the wavefunction you'll find yourself in, once the experiment is complete. That's what a rational agent would do: trust the Born rule. Wallace has tried to derive the Born rule using decision theory, by making certain seemingly simple and acceptable assumptions. For example, if the wavefunction of the universe were to change only by a small amount, your betting strategy should only change by a small amount.

Not everyone is convinced of this approach. The above assumption "would be reasonable with normal physical quantities. [But] is that reasonable when we talk of the wavefunction of the universe? It's such a bizarre thing. How can we get our heads around what this thing really is?" said Wiseman. "It's certainly not just a thing that we are experiencing in the world. It's actually describing us and simultaneously all our possible futures. I'm just not convinced that that problem [of probability] has been solved, despite the work that has been done on it. That's really in my mind the biggest problem with the many worlds interpretation, which has been the problem with it all along. Everett was certainly aware of this problem."

If nothing else, the many worlds interpretation undeniably questions our understanding of the meaning of probabilities in quantum mechanics.

The adherents of many worlds are not the only ones fussing about the meaning of probability. Our final interpretation—at first called Quantum Bayesianism, but now known as QBism—initially got its name from the Bayes rule of probability (named after an eighteenth-century statistician and theologian, Thomas Bayes). Not only is the issue of probability front and center in QBism, but it brings the observer back into the mix, claims that probabilities are subjective (personal to each observer), and throws up questions about what quantum states (the vectors in Hilbert space) say about objective reality. QBism, "rather than relinquishing the idea of reality . . . [says] that reality is more than any third-person perspective can capture."

When Christopher Fuchs was a researcher at the Perimeter Institute in Waterloo, Canada, he and his wife, Kiki, bought an enormous house and refurbished it. The previous owner had been a woman who died in her nineties. The house had a small room where she had watched television and drank and smoked heavily, evidenced by the burn marks on the wooden floor next to the couch. Chris Fuchs thought the room would be perfect for a library with floor to ceiling bookshelves, so Kiki Fuchs designed one. She removed by hand the nicotine-soaked burlap wallpaper, common in houses built in the late nineteenth century, and had the room cleaned up. Then they got a carpenter to build the bookshelves (made of quartersawn oak, because both Chris Fuchs and the carpenter felt that the 1886 house deserved nothing less) and stocked it with Fuchs's favorite books on philosophy, mostly on American pragmatism (by the likes of William James and John Dewey). But there was one section of the library dedicated to the modern American philosopher Daniel Dennett. It wasn't that Fuchs admired Dennett's philosophy—quite the opposite. "The reason Dennett was there was not because I'm a supporter or interested in him in any way, but rather I see him as the enemy," Fuchs told me. "You should know your enemy."

Dennett is a well-known materialist who has long argued that the perceived immateriality of consciousness is an illusion. Fuchs wants to take our conscious experience seriously—a stance he attributes to William James's philosophy. Fuchs has also been heavily influenced by John Wheeler (with whom he studied at the University of Texas at Austin). Wheeler was a staunch advocate of Bohr's vision of quantum mechanics and the Copenhagen interpretation, the strong version of which argues the observer cannot be separated from that which is observed. For Bohr, the observer was some macroscopic experimental setup. Wheeler sometimes went further in his speculations, wondering whether the entirety of existence came down to individual quantum phenomena, each of which was linked to an observer, so that we end up with a universe "built on billions upon billions of elementary quantum phenomena, those elementary acts of observer-participancy."

The alternatives to the Copenhagen interpretation we have seen so far—Bohmian mechanics, collapse theories, many worlds—all remove the observer from the mix (a move that Dennett would likely applaud). Fuchs, however, hitched his wagon to Bohr and Wheeler. He wants to bring the observer back into reckoning. His reading of Wheeler's works in particular led him to thinking about what it means for something to be intrinsically random (which the quantum world is, according to the Copenhagen interpretation, making the probabilities we assign to the outcomes of measurements an objective part of reality). "That led me to think about probability theory," Fuchs told me as we sat at the University of Massachusetts Boston, his new academic home after he moved there from

Waterloo (ironically, Boston is just a few miles away from Dennett's home turf at Tufts University in Medford).

Fuchs's tussle with the meaning of probability in quantum mechanics began in earnest when he was doing his PhD with Carlton Caves at the University of New Mexico in Albuquerque. At the time, Fuchs was a "frequentist"—someone who thinks probabilities are objective measures of the tendencies of things to happen, tendencies that become apparent if you do those things a very large, possibly infinite, number of times. Caves, however, was a Bayesian. In this way of thinking, probability is not an objective property of things. Rather, it's a statement about the person assessing the likelihood of something happening and assigning it a probability: the probability incorporates the idea that the person is uncertain for whatever reason, yet must still make the best decisions possible in light of that uncertainty. Quantum Bayesianism was officially born in 2002, with the first paper by Caves, Fuchs, and Rüdiger Schack. The name proved a mouthful (and besides, the term *Bayesianism* caused controversy, given the many divisions within Bayesian probability theory about the meaning of the term), so Fuchs eventually shortened it to QBism, leaving the *B* to stand for itself. It proved a marketing masterstroke. QBism has a ring to it.

QBism challenged notions about the meaning of the wavefunction. The debate over the status of the wavefunction has been at the heart of all the interpretations we have seen thus far. It can be broadly thought of in two ways: either that the wavefunction represents our knowledge of the quantum system, so it is epistemic, and theories that take this stance are called psi-epistemic; or that the wavefunction is part of reality itself, and theories of this persuasion are called psi-ontic (for ontology).

The Copenhagen interpretation, which doesn't accord any reality to the quantum world beyond what is manifest during observation, is psi-epistemic. The wavefunction contains enough knowledge for us to make probabilistic predictions about the outcomes of experiments. Also, no hidden variables are needed to complete the theory.

There are psi-epistemic models that do not take an anti-realist position, in that they accord a reality to the quantum world, but nonetheless argue that the wavefunction is not a part of this real world; rather, it is about our knowledge of that world. Einstein is thought to have been a proponent of this view of quantum mechanics.

The alternatives to the Copenhagen interpretation examined so far—the de Broglie-Bohm theory, collapse theories, and the many worlds interpretation—are all realist about the wavefunction. They argue for an objective world out there that exists independent of observers. In other words, there is an ontology of the quantum world, and the wavefunction is part of this ontology, making these alternatives psi-ontic.

But in all these cases, the quantum state (given by the wavefunction), whether it's epistemic or ontic, is associated with the quantum system—something that all observers can objectively agree upon. QBism takes a radically different stance. "We'd say: in nature there aren't any things called quantum states," Fuchs said. "They just aren't out there."

QBism is definitely psi-epistemic, but the wavefunction is associated with each individual observer studying a quantum system, not with the quantum system. So if I'm making a quantum measurement, the wavefunction I use for the quantum system encodes my expectations for the consequences of the action I'm about to take on it. These expectations are dictated by my beliefs about the system.

So take a photon going through a beam splitter. If you, the agent, had no idea of what a beam splitter does, then you might associate a wavefunction to the photon as it goes past the beam splitter that encodes your uncertainty. The wavefunction is a linear combination of two components (the transmitted and the reflected paths). Let's say you assign a probability of 1/3 for finding the photon in detector D1 and of 2/3 for finding it in detector D2. This is the same as assigning amplitudes of $1/\sqrt{3}$ and $\sqrt{2}/\sqrt{3}$ to the reflected and transmitted parts of the wavefunction, so that squaring the amplitudes gives you the respective probabilities.

But experience—say of doing the experiment over and over again or understanding the physics of the beam splitter or reading textbooks or talking with your colleagues—would tell you that these probabilities are wrong. You'd update the wavefunction until it

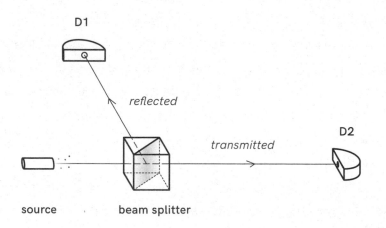

D1

reflected

D2

transmitted

source   beam splitter

accurately represents your belief that the photon will go to D1 half the time and D2 half the time. This idea can be extended to the full Mach-Zehnder interferometer and, by extension, to the double slit, though the argumentation gets more involved. Regardless of the complexity of the system, the key idea here is that the probabilities you assign to the outcomes of experiments are contingent upon your personal set of beliefs about what might happen.

"All of this is personal Bayesian probability," said Fuchs. "The Bayesian notion is that you write down a probability assignment as a measure of what you can't predict. You might write down a probability assignment just because you don't have all the facts. [It's] not an objective feature of things, it's rather a statement about the person making the [prediction]."

Some physicists have argued that QBism is simply Copenhagen in sheep's clothing. But Fuchs vehemently disagrees. He points out that in the Copenhagen interpretation, the wavefunction is associated with the quantum system being studied, and removing the observer doesn't remove the wavefunction: it still exists, independent of the observer, as an objective, epistemic statement about the system. Not so in QBism. Remove the observer and there is no quantum state, no wavefunction to talk of. Moreover, Copenhagen is anti-realist. QBism is not, according to Fuchs. It does not deny that there is a real world out there. All it does is state, unequivocally, that the quantum states in the formalism are not about the real world but about our beliefs about the real world. They are subjective, not objective.

What does all this achieve, however? For one, the whole issue of the collapse of the wavefunction becomes, well, a nonissue. There is nothing physical that is collapsing. All that happens is that you update your beliefs about the world: the wavefunction, which quantifies your expectations, changes. Nothing physical happened (note that the same argument can be made for any psi-epistemic theory—that the collapse has nothing to do with the physical system, but rather, it's about the change in our knowledge of the system).

"For QBism, you don't need a physical story anymore," said Fuchs about the need to explain collapse. "Instead you say this: I took an action that led to a consequence, and because of the consequence I believe new things. The things I believe are captured by this mathematical symbol, ψ. Because I believe new things, instantaneously, upon the new experience, this mathematical object changes instantaneously."

As far as interpretations or alternative theories of the quantum world go, QBism is one of the newest kids on the block, and it goes against the grain for most physicists, who shrink from the idea of personalizing science. For a while, except for Fuchs and Shack, there were few takers. But QBism got a boost when David Mermin, a highly regarded solid state physicist at Cornell University in Ithaca, New York, came on board.

"What really appealed to me about QBism was that it gave a context in which Copenhagen made more sense, and gave an explanation of why Copenhagen was so hard to grasp," Mermin told me when I met him at his office in Ithaca on a blustery, bitingly cold winter's day. "Because what everybody was doing was what scientists had been taught to do almost forever, which was to construct an understanding of the external world that made no reference whatsoever to the people who are trying to understand it. And a lot of the clumsy, awkward things about Copenhagen involved trying to objectify things that aren't objective, that are subjective and personal."

QBism lays itself open to charges of solipsism, which is the argument that the only thing that is real is what I experience. According to Mermin, there is a fallacy in such arguments. We have language to communicate with each other about our private experiences—including the language of science and mathematics—and this makes our subjective experience a shared reality.

Nonetheless, there is nothing like an objective third-person view of reality in QBism. This has implications for questions about whether the universe is local or nonlocal or whether there is a quantum-classical divide—the other axes along which we can sort out the various interpretations. Take nonlocality. Copenhagen argues that the quantum world is nonlocal but provides no explanations for why it might be so. It just is. The de Broglie-Bohm theory resorts to nonlocal hidden variables to explain nonlocality. Collapse theories are nonlocal, in that they take the wavefunction seriously—and the collapse of the wavefunction (whether it happens stochastically as in GRW or because of gravity as in the Diósi-Penrose theories) is a nonlocal event. And according to some proponents of the many worlds interpretation, the universe is local. QBism says so too. And both of them resort to similar arguments to argue why.

Recall Alain Aspect's experiment about Alice and Bob doing measurements on entangled photons and finding correlations that imply instantaneous action at a distance between, say, Alice's measurements and Bob's photons, or vice versa. So if Alice and Bob make measurements, and their measurements have definite outcomes, then when analyzed from a third-person perspective, these measurement outcomes are correlated in ways that cannot be explained without the assumption of nonlocality. But the third-person perspective doesn't make sense in the many worlds interpretation. In his book *The Emergent Multiverse*, David Wallace argues that when it comes to many worlds, "from the third-person perspective from which Bell's theorem is normally discussed, no experiment has any unique definite outcome at all."

Wallace explains: "From the perspective of a given experimenter, of course, her experiment *does* have a unique, definite outcome, even in the Everett interpretation. But Bell's theorem requires more: it requires that from her perspective, her distant colleague's experiment also has a definite outcome. This is not the case in Everettian quantum mechanics—not, at any rate, until that distant experiment enters her past light cone." Meaning that the instant at which one can talk about measurements by Alice and Bob in the same breath is when one gets access to the other's world—which cannot happen faster than the speed of light.

On the whiteboard in his office in Boston, Fuchs drew cartoons for Alice and Bob to explain why QBism takes a somewhat similar stance. "This is a bit like many worlds, which I'll be honest about," he said. In both QBism and many worlds, from Alice's perspective, there's no click on Bob's detector and vice versa. Bell's analysis, however, insists that there be a click on both sides, as seen from a third-person perspective. This isn't possible in QBism. It's only after Alice walks over to Bob—which cannot happen faster than the speed of light—and Bob's results become part of her experience that she can update her beliefs about what has happened. But until then, there is no notion of correlations between the results obtained by Alice and Bob. "Quantum mechanics, in the QBist interpretation, cannot assign correlations, spooky or otherwise, to space-like separated events, since they cannot be experienced by any single agent. Quantum mechanics is thus *explicitly* local in the QBist interpretation. And that's all there is to it."

There's a similar dismissal of the idea of the quantum-classical divide in QBism. In the Copenhagen interpretation, there is a divide that exists by diktat. There are some things that are quantum and others that are classical, but without any solid explanation for why that might be so. Explanations using decoherence get us partway to understanding why quantum states might end up as classical, but they don't complete the job. In the de Broglie-Bohm theory, there is no divide. There is always a fact of the matter as to where the particles that make up any object are, however big or small the object. Collapse theories argue for an emergent divide that is brought about by the stochastic process of collapse itself. The many worlds interpretation does not distinguish between the classical and the quantum—the wavefunction is all there is, evolving, forever. QBism, on the other hand, asks us to rethink the very notions of what we mean by the quantum and the classical, given that these terms are usually talked about from an impersonal, objective third-person perspective. "Science is about the interface between the experience of any particular person and the subset of the world that is external to that particular user . . . It is central to the QBist understanding of science," wrote David Mermin in his essay "Why QBism Is Not the Copenhagen Interpretation and What John Bell Might Have Thought of It." So, in QBism, what one thinks of (and "one" here means a particular person) as classical or quantum has simply to do with one's beliefs about the world outside.

If all this makes our heads reel, we should rest assured that we are not the only ones suffering. Physicists who are deeply immersed in these questions are not immune to being flummoxed. There are experts on Bohmian mechanics who profess to being clueless about QBism, QBists who think collapse theories are misguided, collapse theorists who claim the many worlds theory is extravagant nonsense, and advocates of many worlds who dismiss Bohmian mechanics as unnecessarily contrived. And, of course, all those who work on alternative interpretations of quantum mechanics think Copenhagen should be consigned to the dustbin of history. And the Copenhagen folks, well, they are yet to be decisively knocked down from their somewhat lofty perch.

Some young minds—à la Heisenberg when he was twenty-four—might cut through this clutter. There's insight in a comment that Anton Zeilinger made to Fuchs, after Fuchs had given a talk on QBism that wasn't terribly well received in Fuchs's own estimation. Fuchs thought that the veteran Alain Aspect, who was in the audience, had written him off "as a nutcase." Even Mermin, a well-wisher, walked up to Fuchs and said, "We need to talk. That was the worst talk you've ever given." Zeilinger said, "Great talk!" to which Mermin responded, "No, it wasn't!" Fuchs recalled (in his writings) that Zeilinger looked past Mermin and addressed him directly: "You know what I would do when I was young, being dismissed by the old professors on the front rows of the seminar? I would not look to them as I spoke, but rather to the back rows where the young students were sitting. They were the ones ready to hear something new."

It might well take someone new, young, and unbiased to make incontrovertible sense of the quantum world. I met physicists who are deeply convinced that they are on the correct path, as they must be to summon the energy needed to devote an entire lifetime to the pursuit of the nature of reality. I met physicists who remain dissatisfied with the status quo, uncommitted to any one path, as they should be to discern any cracks in the foundations of quantum mechanics. Surely, all interpretations and formalisms can't be simultaneously correct. Maybe one of them is, maybe none are. Or, tantalizingly, maybe they are all touching the truth in their own way and giving us glimpses of a deeper reality. If so, the cracks will let the light through, and we'll be better able to tell whether it goes through two doors at once. Or not.

Epilogue

WAYS OF LOOKING
AT THE SAME THING?

During the late 1970s and early '80s, Werner Erhard, the founder of est, organized a series of physics conferences, using the wealth from his self-help empire to indulge his fascination for physics. "The est foundation's physics conferences attracted star after star of the physics firmament," wrote David Kaiser in his book *How the Hippies Saved Physics: Science, Counterculture, and the Quantum Revival.* One of these stars was Leonard Susskind, a theoretical physicist at Stanford University. One evening, Susskind was having dinner with Richard Feynman and Sidney Coleman at Erhard's home in San Francisco. Erhard had also invited two young philosophers to the dinner. "They were spouting all sorts of philosophical verbiage, academic style philosophical verbiage . . . which it was clear that Feynman had no patience with and he took them apart. It was cruel. I don't know how to describe what he did to them. With simple words, he took a pin and punctured their balloon in a way that you might call ugly, but the saving grace is that they were totally enchanted by him," Susskind told me.

But despite his distaste for bloviating philosophers, Feynman "was possibly the most philosophical of all the physicists I ever knew," said Susskind.

This side of Feynman was clearly evident during his lectures at Cornell. In one talk, he asked his audience to consider two theories, A and B, which have different takes on the nature of reality but which are mathematically equivalent, make the same empirical predictions, and are impossible to tell apart experimentally (he could have been talking about the Copenhagen interpretation and Bohmian mechanics, but he wasn't—he was making a general point). Feynman argued that it's important to understand that the philosophies behind A and B can lead us in different directions even if they are indistinguishable at some stage of the scientific process.

"In order to get new theories, these two things are very far from equivalent. Because one gives a man different ideas than the other," said Feynman.

For example, it might be possible to make a tiny tweak to A that isn't possible with B. In which case, A can lead to a very different theory after the change. "In other words, although they are identical before they are changed, there are certain ways of changing one which look natural, which don't look natural in the other. Therefore, psychologically we must keep all the theories in our head," said Feynman. "And every theoretical physicist that's any good knows six or seven different theoretical representations for exactly the same physics and knows that they are all equivalent and that nobody is ever going to be able to decide which one is right at that level . . . but he keeps them in his head, hoping that they'll give him different ideas."

In Brisbane, Australia, Howard Wiseman tries to do exactly that: keep the different interpretations of quantum mechanics in mind, and see what emerges. One obvious intuition is that these theories and interpretations are each shining a light on a different aspect of the same reality. "A lot of progress in philosophy of science has been made by showing that things which were thought of as being separate are actually just different ways of looking at the same thing," Wiseman told me.

This approach, when applied to quantum mechanics, is providing some surprising insights. Take collapse theories and a hidden variable theory like Bohmian mechanics—two very different views of the nature of reality. In Bohmian mechanics, if you consider a two-particle system, the wavefunction is a function of two variables, the position of particle A and the position of particle B; these particles also have actual positions, the hidden variables in the theory. Now, hypothetically, if you knew the exact position of particle A (which in practice you cannot, but let's go with the argument) you can plug that into the equations and the wavefunction reduces—or effectively collapses—to a function of one variable, the position of particle B.

This inspired Wiseman to think of collapse theories—in which the wavefunction stochastically collapses at some rate at different points in spacetime—as theories in which the wavefunction is entangled with some other large system with hidden variables that we are unaware of. Changes to the values of these hidden variables, unbeknownst to us, can influence the wavefunction of the system we are studying—and it can seem like the wavefunction is subject to stochastic collapses. In this way of thinking about collapse theories, they become hidden variable theories, except that the variables are, well, truly hidden—and we are privy only to their effects.

Wiseman has also found ways of connecting Bohmian mechanics to the idea of many worlds. Take a single particle going through a double slit in Bohmian mechanics. If you knew its exact initial position and velocity, you could predict its exact trajectory through the apparatus. But in order to tally with the probabilistic predictions of quantum mechanics, Bohmian mechanics adds a dash of uncertainty to our knowledge about the initial state of the particle. Its starting position is given by a probability field, meaning the particle could be in one of many locations, with a different probability for each starting position. It's like imagining a virtual ensemble of particles whose starting positions are dictated by this probability field, given by the modulus-squared of the wavefunction. The real particle is in one of those positions, we just don't know which. Now, as this virtual ensemble evolves through the double slit, Bohmian mechanics gives us a virtual ensemble of trajectories, but, of course, only one of them is real—and for one run of the experiment, we'll see the consequence of one such trajectory. We'll find the particle landing somewhere on a screen, an outcome to which we could have assigned only a certain probability. And if you do this experiment over and over again, with the same virtual ensemble of particles, you'll get a slew of trajectories, which taken together form an interference pattern.

When contemplating this virtual ensemble, Wiseman wondered: what if this virtual ensemble is real, in that every particle exists, but in a different world? Each particle is influenced by those in its local surroundings (some regions are more dense with particles than others). Then the way each particle moves is dictated by interactions with its immediate neighbors, like starlings in a murmuration. Crucially, you don't need a wavefunction to determine how any given particle moves. "When that idea occurred to me, it was like, wow!" Wiseman said.

He and his colleagues Dirk-André Deckert and Michael Hall postulated a particular many-world force law for this situation, then did simulations. First, they used Bohmian mechanics—wavefunction, hidden variables, and all—to plot trajectories of particles going through a double slit. Next, they treated each particle trajectory as being the outcome of its interactions with other particles that exist in other worlds, without resorting to the mathematics of the evolution of the wavefunction. The results they obtained are eerily similar.

"That's just about the theory for one particle," said Wiseman. "Of course, the universe is not one particle."

In a many-particle system, while each particle exists in 3-D space, the wavefunction exists in configuration space. Any given distribution of particles in 3-D space corresponds to one point in the configuration space. This one point represents a world. In Bohmian mechanics, our knowledge about the initial configuration of particles is subject to uncertainty—and this uncertainty corresponds to a probabilistic distribution of points in configuration space, or a virtual ensemble of worlds. Bohmian mechanics says there is one real world whose evolution can be described by the evolution of the wavefunction, subject to this initial uncertainty. Just as in the case of the single particle in 3-D space being influenced

DE BROGLIE–BOHM THEORY

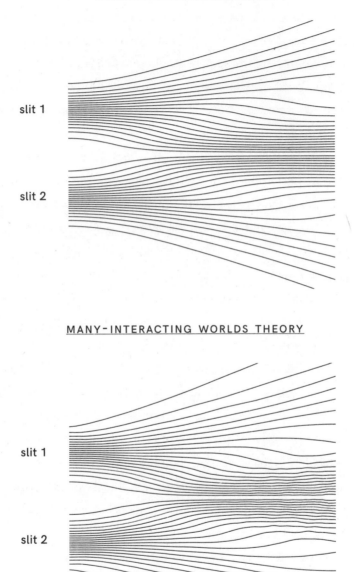

slit 1

slit 2

MANY-INTERACTING WORLDS THEORY

slit 1

slit 2

by other particles in other worlds, Wiseman's argument is that you can treat this virtual ensemble of points in configuration space as a collection of real worlds. Any given world interacts with the other worlds in configuration space, and this interaction is local, with nearby worlds influencing each other more than distant worlds.

This has enormous consequences. For example, a local interaction in configuration space can appear nonlocal in 3-D space. "So that's where the nonlocality of quantum mechanics would come from," said Wiseman. He's dubbed this still-nascent idea "many interacting worlds" (to distinguish it from the Everettian many worlds idea). It's an example of how the behavior of the quantum mechanical world emerges from the dynamics of a deeper reality. But he's certainly not claiming that this is what's happening in the quantum world. The exercise is to make the point that there are myriad ways of explaining quantum phenomena, some of which are on stronger mathematical footing than others, and each has its own set of problems, whether it's the measurement problem in the Copenhagen interpretation, or the problem of potential conflicts with special relativity in Bohmian mechanics (not to mention the distaste for hidden variables in the eyes of some), or the ad hoc nature of stochastic collapse in collapse theories, or the problem of explaining probability in Everett's many worlds.

And depending on which axis you choose for analyzing the quantum world, the theories and interpretations fall into different bins, making for strange bedfellows.

Take determinism. The de Broglie-Bohm theory, the Everettian many worlds interpretation, and Wiseman's many interacting worlds are deterministic; Copenhagen and collapse theories are not.

QBism doesn't say anything about whether or not the real world is deterministic.

What about realism? Well, de Broglie-Bohm, collapse, many worlds, and many interacting worlds are all realist. Copenhagen is not. QBism is realist, with the caveat that the wavefunction is not about this reality.

What about claims that there is nothing but the wavefunction? Many worlds and collapse theories say yes. De Broglie-Bohm says no (because the theory has hidden variables besides). There is no wavefunction in many interacting worlds. In Copenhagen, the wavefunction represents the quantum world, but there is the classical world to contend with. In QBism, the status of the wavefunction is entirely subjective (it's personal to one observer).

Then there's the whole issue of locality versus nonlocality. From the point of view of our 3-D world, the theories of de Broglie-Bohm, collapse, and many interacting worlds are all nonlocal. There's some dispute about the status of the Everettian many worlds interpretation in this regard, but opinions tend toward it being local. Copenhagen is obscure: if you take the wavefunction to be representative of something out there, then it's nonlocal, else you can dismiss concerns of nonlocality by saying that all one does is make measurements, compare them to results of other experiments, find correlations, and that's that (there's no attempt to explain the cause of the correlations). QBism, as we saw earlier, dismisses nonlocality.

There are finer distinctions to be made, but the message is clear: there's no way to classify these theories in a consistent manner. It's a strong clue that our understanding of the quantum world is still up for grabs. And it's very likely that further attempts at clarification will involve the double-slit experiment in some form or the other.

And nowhere does this become more apparent than in experiments being done to verify one of the key assumptions of quantum mechanics: the Born rule. As one prominent theorist has said, "If Born's rule fails, everything goes to hell." All of the various mathematical formalisms for explaining the quantum world, ultimately, are designed to answer why, when we do experiments like the double slit, we get the outcomes we do. A photon goes through the double slit, its wavefunction splits, evolves, recombines, and so on. Eventually, the wavefunction can be written down as a linear combination of different wavefunctions (one for each path the photon takes), each evolving according to the rules of the Schrödinger equation. The photon is said to be in a superposition of taking all possible paths. The Born rule says that the probability of finding the photon at any given location is given by the modulus-squared of the value of the wavefunction at that location. But "the Born Rule is conjecture," said Urbasi Sinha, formerly of the Institute for Quantum Computing at Waterloo, Canada, and now of the Raman Research Institute in Bengaluru, India. "There is no formal proof of the Born Rule."

Of course, there are many quantum mechanical phenomena that agree with theoretical predictions to astonishing precision, but these predictions all assume the validity of the Born rule. Now, Sinha and her colleagues are trying to directly test the Born rule—using, what else, the double-slit (or sometimes the triple-slit experiment, which makes some measurements sharper but is conceptually identical to doing the experiment with two slits). Take a photon going through the double slit. According to Feynman's path integral approach, to calculate the probability of finding the photon at, say, the center of the screen, you have to consider the classical paths (which go through one slit or the other) and nonclassical paths, such as one that starts off going through one slit, then instantly turns toward the other slit, and then goes toward the screen.

Sinha's team calculated the expected intensity of light at the center of the screen, given the most dominant classical and nonclassical paths the photon can take through the apparatus. The next step was to figure out the expected change in intensity if the nonclassical path is blocked (which can be done by placing a baffle through one slit—this prevents the photon from taking its unusual slit-hugging route. The baffle is thin, so there is still room on either side of it for a photon taking the classical path to go through the slit). The measured intensity is sensitive to the exact formulation of the Born rule. Is the probability equal to the square of the amplitude of the wavefunction? Or is it equal to the amplitude raised to some number $2+\delta$, where δ represents a tiny deviation? Many teams besides Sinha's are asking such questions. "Even a small deviation will change many things," Sinha told me.

While the Born rule has held up so far to a certain level of precision, the experimentalists are tightening the screws. If they can show that the Born rule needs tweaking, it will create an opening, giving theorists essential clues on how to proceed toward the correct quantum mechanical view of nature. The experiments also highlight that the double slit, a simple contraption if ever there was one, continues to conceal some central principle that animates reality.

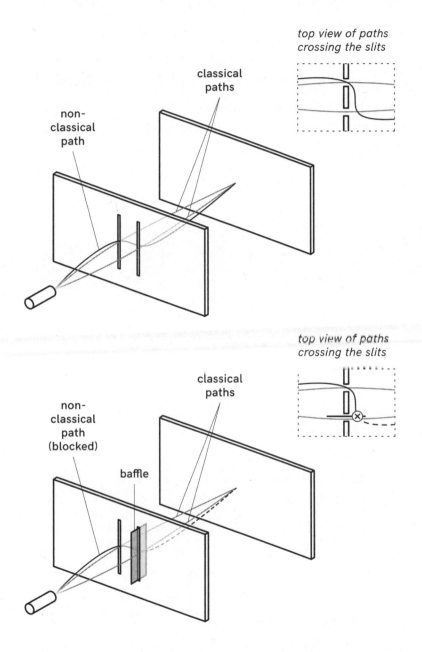

top view of paths
crossing the slits

classical
paths

non-
classical
path

top view of paths
crossing the slits

classical
paths

non-
classical
path
(blocked)

baffle

As Feynman put it in his Cornell lecture: "Any . . . situation in quantum mechanics, it turns out, can always be explained afterwards by saying 'You remember the case of the experiment with the two holes?'"

Physics has yet to complete its passage through the double-slit experiment. The case remains unsolved.

NOTES

vii **"Allow me to express now"**: Carl C. Gaither and Alma E. Cavazos Gaither, eds., *Gaither's Dictionary of Scientific Quotations* (New York: Springer, 2008), 502.

5 **"There is nothing more surreal"**: Siri Hustvedt, "The Drama of Perception: Looking at Morandi," *Yale Review* 97, no. 4 (Oct 2009): 20–30.

5 **But in November 1964**: http://www.cornell.edu/video/playlist/richard-feynman-messenger-lectures.

6 **"It's odd, but in the infrequent occasions"**: Feynman Messenger Lectures, Lecture 1, "Law of Gravitation," http://www.cornell.edu/video/richard-feynman-messenger-lecture-1-law-of-gravitation.

6 **"Then we see unexpected things"**: Feynman Messenger Lectures, Lecture 6, "Probability and Uncertainty: The Quantum Mechanical View of Nature," http://www.cornell.edu/video/richard-feynman-messenger-lecture-6-probability-uncertainty-quantum-mechanical-view-nature.

6 **"They behave in their own inimitable way"**: Ibid.

7 **"That is, they're both screwy"**: Ibid.

7 **"But the difficulty"**: Ibid.

7 **"one experiment which has been designed"**: Ibid.

11 **let's orient the device**: The British physicist Jim Al-Khalili has used the same idea to demonstrate the double-slit experiment done with particles, https://youtube/A9tKncAdlHQ?t=125.

13 **"The Last Man Who Knew Everything"**: Andrew Robinson, *The Last Man Who Knew Everything* (London: OneWorld, 2007).

13 **"doctor of physic, surgery, and midwifery"**: Ibid., 51.

13 **"The experiments I am about to relate"**: Thomas Young, "The Bakerian Lecture: Experiments and Calculations Relative to Physical Optics," *Philosophical Transactions of the Royal Society of London* 94 (1804): 1–16.

13 **"I made a small hole in a window-shutter"**: Ibid.

14 **"I brought into the sunbeam"**: Ibid.

15 **Young saw such optical interference fringes**: Ibid.

15 **Henry Brougham**: https://www.britannica.com/biography/Henry-Peter-Brougham-1st-Baron-Brougham-and-Vaux.

15 **"destitute of every species"**: Whipple Museum of the History of Science, http://www.sites.hps.cam.ac.uk/whipple/explore/models/wavemachines/thomasyoung/#ref_2.

23 **"The idea of an objective real world"**: Werner Heisenberg, *Physics and Philosophy* (London: Penguin Books, 2000), 83.

23 **He elucidated his laws**: https://www.aps.org/publications/apsnews/200007/history.cfm.

24 **"When Light reaches us from the sun"**: Louis de Broglie, *Matter and Light: The New Physics*, trans. W. H. Johnston (New York: W. W. Norton & Co., 1939), 27.

24 **He presented these ideas on December 8, 1864**: J. Clerk Maxwell, "A Dynamical Theory of the Electromagnetic Field," *Philosophical Transactions of the Royal Society of London* 155 (1865): 459–512.

25 **In 1879, the Prussian Academy of Sciences**: D. Baird, R. I. Hughes, and A. Nordmann, eds., *Heinrich Hertz: Classical Physicist, Modern Philosopher* (Dordrecht, NL: Springer Science, 1998), 49.

25 **"But in spite of having abandoned"**: Ibid.

25 **"It is of no use whatsoever"**: Andrew Norton, ed., *Dynamic Fields and Waves* (Bristol: CRC Press, 2000), 83.

27 **"To be sure, it is a discovery"**: Joseph F. Mulligan, "Heinrich Hertz and Philipp Lenard: Two Distinguished Physicists, Two Disparate Men," *Physics in Perspective* 1, no. 4 (Dec 1999): 345–66.

28 **"A chronic, and painful, disease"**: "Heinrich Hertz," editorial in *Nature* 49, no. 1264 (Jan 18, 1894): 265.

28 **"Heinrich Hertz seemed to be predestined"**: Mulligan, "Heinrich Hertz and Philipp Lenard."

28 **was a scientific curiosity**: https://history.aip.org/history/exhibits/electron/jjrays.htm.

29 **"At first there were very few":** http://history.aip.org/exhibits/electron /jjelectr.htm.

29 **His experiments clearly showed that:** Mulligan, "Heinrich Hertz and Philipp Lenard."

30 **While Einstein did not fully embrace Planck's ideas:** Abraham Pais, "Einstein and the Quantum Theory," *Reviews of Modern Physics* 51, no. 4 (Oct 1979): 863–914.

32 **"Entrance is forbidden to Jews":** Mulligan, "Heinrich Hertz and Philipp Lenard."

32 **"Einstein was the embodiment":** Philip Ball, "How 2 Pro-Nazi Nobelists Attacked Einstein's 'Jewish Science' " excerpt, February 13, 2015, https://www .scientificamerican.com/article/how-2-pro-nazi-nobelists-attacked -einstein-s-jewish-science-excerpt1/.

33 **Thomson argued that there should be blurry fringes:** George K. Batchelor, *The Life and Legacy of G. I. Taylor* (Cambridge: Cambridge University Press, 1996), 40.

34 **"I chose that project for reasons":** Ibid.

34 **To create a single slit:** Ibid., 41.

34 **"I had, I think rather skillfully, arranged":** Ibid.

34 **Taylor reportedly went away sailing:** Sidney Perkowitz, *Slow Light: Invisibility, Teleportation, and Other Mysteries of Light* (London: Imperial College Press, 2011), 68.

34 **After that three-month-long exposure:** George K. Batchelor, *The Life and Legacy of G. I. Taylor*, 41.

37 **it'd take him a decade or so more:** Gösta Ekspong, "The Dual Nature of Light as Reflected in the Nobel Archives," https://www.nobelprize.org/nobel_prizes /themes/physics/ekspong/.

38 **"That he might sometimes have overshot":** Walter Isaacson, *Einstein: His Life and Universe* (New York: Simon & Schuster, 2007), 100.

39 **The moment, captured in a now-iconic photograph:** Participants of the Fifth Solvay Congress, https://home.cern/images/2014/01/participants-5th -solvay-congress.

40 **Bohr first met Heisenberg:** Jagdish Mehra, *Einstein, Physics and Reality* (Singapore: World Scientific, 1999), 94.

40 **He also invited Heisenberg to Copenhagen:** Gino Segrè, *Faust in Copenhagen: A Struggle for the Soul of Physics* (New York: Viking Penguin, 2007), 116.

40 **"perhaps it would be possible one day":** Jagdish Mehra, *Golden Age of Theoretical Physics, vol. 2* (Singapore: World Scientific, 2001), 648.

41 **There, between long walks and contemplating:** Ibid., 650.

41 **"It was almost three o'clock in the morning":** Ibid., 651.

41 **"I thought the whole day":** Ibid., 652.

45 **"discouraged, if not repelled":** Ibid., 840.

45 **"A few days ago I read":** Walter Moore, *Schrödinger: Life and Thought* (Cambridge: Cambridge University Press, 2015), 192.

45 **"A few days before Christmas":** Dick Teresi, "The Lone Ranger of Quantum Mechanics," review of Walter Moore's *Schrödinger: Life and Thought,* January 7, 1990, http://www.nytimes.com/1990/01/07/books/the-lone-ranger-of-quantum -mechanics.html.

48 **"The motion of particles follows probability":** Abraham Pais, "Max Born's Statistical Interpretation of Quantum Mechanics," *Science* 218 (Dec 17, 1982), 1193–98.

49 **He wrote to Pauli, complaining:** Moore, *Schrödinger,* 221.

49 *Züricher Lokalaberglauben:* Ibid.

49 **"Don't take it as a personal unfriendliness":** Ibid.

49 **"The discussion between Bohr and Schrödinger":** Ibid., 226.

50 **"Schrödinger was a 'visualizer'":** Ibid., 228.

50 **"trust in the newly developed":** Stefan Rozental, ed., *Niels Bohr: His Life and Work as Seen by His Friends and Colleagues* (Amsterdam: North-Holland Publishing, 1967), 104.

50 **"Like a chemist who tries to concentrate":** Jørgen Kalckar, ed., *Niels Bohr Collected Works, vol. 6* (Amsterdam: North-Holland, 1985), 15.

51 **"I went for a walk":** Rozental, *Niels Bohr,* 105.

52 **"Dear Mr. Bohr":** Léon Rosenfeld and J. Rud Nielsen, eds., *Niels Bohr Collected Works, vol. 3* (Amsterdam: North-Holland, 1976), 22.

53 **"To meet you":** Ibid.

54 **"However, with all the participants":** Manjit Kumar, *Quantum: Einstein, Bohr, and the Great Debate about the Nature of Reality* (New York: Norton, 2011), 273.

56 **physicists since then have reimagined:** The illustration of the recoiling double slit is inspired by a drawing in P. Bertet et al., "A Complementarity Experiment with an Interferometer at the Quantum-Classical Boundary," *Nature* 411 (May 10, 2001): 166–70.

58 **"But the electron must be somewhere":** Jorrit de Boer, Erik Dal, and Ole Ulfbeck, eds., *The Lesson of Quantum Theory* (Amsterdam: North-Holland, 1986), 17.

59 **"The electron, as it leaves the atom":** Arthur Eddington, *The Nature of the Physical World* (New York: Macmillan Company, 1929), 199.

59 **"It is fair to state that we are not":** Stefan Hell, Nobel Banquet Speech, December 10, 2014, https://www.nobelprize.org/nobel_prizes/chemistry /laureates/2014/hell-speech_en.html.

59 **"Now, ladies and gentlemen":** Ibid.

61 **a decade-old paper by John Bell:** J.S. Bell, "On the Einstein Podolsky Rosen paradox," *Physics Physique Fizika* 1, no. 2 (Nov 1, 1964): 195–200.

63 **Taylor used something called a coincidence detector:** William M. Honig, David W. Kraft, and Emilio Panarella, eds., *Quantum Uncertainties: Recent and Future Experiments and Interpretations* (New York: Plenum Press, 1987), 339.

63 **A rough-and-ready calculation:** The calculation appears in Giancarlo Ghirardi, *Sneaking a Look at God's Cards: Unraveling the Mysteries of Quantum Mechanics* (Princeton: Princeton University Press, 2005), 16.

64 **"This experiment has never been done":** Richard P. Feynman, Robert B. Leighton, and Matthew Sands, *The Feynman Lectures on Physics, vol. 1, New Millennium Edition* (New York: Basic Books, 2011), 37–5.

65 **Möllenstedt noticed that when the tungsten:** Edgar Völkl, Lawrence F. Allard, and David C. Joy, eds., *Introduction to Electron Holography* (New York: Springer Science, 1999), 3.

65 **"kept a collection of spiders":** Robert Crease, *The Prism and the Pendulum: The Ten Most Beautiful Experiments in Science* (New York: Random House, 2004), 197.

65 **the team did not see any fringes at first:** Völkl, Allard, and Joy, eds., *Introduction to Electron Holography*, 5.

66 **"Thomas Young had produced":** Ibid.

66 **"It was . . . a great pleasure to see":** Ibid., 7.

67 **In 1974, Italian physicists:** Pier Giorgio Merli, GianFranco Missiroli, and Giulio Pozzi, "On the Statistical Aspect of Electron Interference Phenomena," *American Journal of Physics* 44, no. 306 (1976): 306–7.

67 **the movie even won an award:** https://www.bo.imm.cnr.it/users/lulli /downintel/electroninterfea.html.

67 **In 1989, Akira Tonomura and colleagues:** A. Tonomura et al. "Demonstration of Single Electron Buildup of an Interference Pattern," *American Journal of Physics* 57, no. 117 (1989): 117–20.

68 **"We believe that we carried out":** Letter to editor, "The Double-Slit Experiment with Single Electrons," *Physics World* (May 2003): 20.

74 **Such an excited atom falls back:** Alain Aspect, Philippe Grangier, and Gérard Roger, "Experimental Tests of Realistic Local Theories via Bell's Theorem," *Physical Review Letters* 47, no. 7 (Aug 17, 1981): 460–63.

79 **"just a name for something":** David Albert, *Quantum Mechanics and Experience* (Cambridge, MA: Harvard University Press, 1994), 11.

80 **"the *most* unsettling story perhaps":** Ibid., 1.

82 **Einstein imagined some gunpowder:** Arthur Fine, *The Shaky Game: Einstein, Realism and the Quantum Theory* (Chicago: University of Chicago Press, 1986), 78.

82 **"Through no art of interpretation":** Ibid.

82 **"I am long past the stage":** Ibid., 82.

83 **"A cat is shut up in a steel chamber":** Moore, *Schrödinger*, 308.

84 **a particle interferes with itself:** Paul Dirac, *The Principles of Quantum Mechanics* (Oxford: OUP, 1958), 9.

85 **"great smoky dragon":** Warner A. Miller and John A. Wheeler, "Delayed-Choice Experiments and Bohr's Elementary Quantum Phenomenon," S. Kamefuchi et al., eds., *Proceedings of the International Symposium on Foundations of Quantum Mechanics* (Tokyo: Physical Society of Japan 1984), 140–52.

86 **"What the dragon does":** Ibid.

88 **"One decides whether the photon":** John Wheeler and Wojciech Zurek, eds., *Quantum Theory and Measurement* (Princeton: Princeton University Press, 1983), 183.

89 **an interferometer with arm lengths of 48 meters:** Vincent Jacques et al., "Experimental Realization of Wheeler's Delayed-Choice Gedanken Experiment," *Science* 315, no. 5814 (Feb 16, 2007): 966–68.

91 **Einstein's most cited paper:** Dennis Overbye, "Quantum Trickery: Testing Einstein's Strangest Theory," *New York Times,* December 27, 2005, http:// www.nytimes.com/2005/12/27/science/quantum-trickery-testing-einsteins -strangest-theory.html.

93 **"Nonlocality forces us":** Nicolas Gisin, *Quantum Chance: Nonlocality, Teleportation and Other Quantum Marvels* (Cham, Switzerland: Springer, 2014), 32.

94 **"first apologized for not having":** Andrew Whitaker, *Einstein, Bohr and the Quantum Dilemma: From Quantum Theory to Quantum Information* (Cambridge: Cambridge University Press, 2006), 203.

95 **Einstein made the point that this localization:** Ibid.

95 **EINSTEIN ATTACKS QUANTUM THEORY:** Kelly Devine Thomas, "The Advent and Fallout of EPR," *IAS: The Institute Letter* (Fall 2013): 13.

96 **"I deprecate advance publication":** Ibid.

96 **"If *The New York Times* is the secular press":** David Mermin, Oppenheimer Lecture, University of California, Berkeley, March 17, 2008, https://youtube /ta09 WXiUqcQ?t=833.

96 **elegant four-page-long paper:** Albert Einstein, Boris Podolsky, and Nathan Rosen, "Can Quantum-Mechanical Description of Physical Reality Be Considered Complete?" *Physical Review* 47 (May 15, 1935): 777–80.

96 **teatime conversation between Einstein and Rosen:** Thomas, "Advent and Fallout of EPR."

98 **Sixteen years later, in 1951:** David Bohm, *Quantum Theory* (New York: Dover Publications, 1989), 611.

99 *"If, without in any way":* Einstein, Podolsky, and Rosen, "Quantum-Mechanical Description."

99 **"a piece of historical silliness":** Fine, *Shaky Game*, 57.

100 **"the first and only female doctoral student":** Elise Crull and Guido Bacciagaluppi, eds., *Grete Hermann—Between Physics and Philosophy* (Dordrecht: Springer, 2016), 4.

100 **"A thorough examination of the proof of von Neumann reveals":** Harald Atmanspacher and Christopher A. Fuchs, eds., *The Pauli-Jung Conjecture: And Its Impact Today* (Exeter, UK: Imprint Academic, 2014), ebook.

100 **"Why should we believe in that?":** Ibid.

101 **"would rather be a cobbler":** Isaacson, *Einstein*, 324.

101 **"If that had something to do with it":** Crull and Bacciagaluppi, *Grete Hermann*, 184.

101 **"In 1952, I saw the impossible done":** Olival Freire Jr., *The Quantum Dissidents: Rebuilding the Foundations of Quantum Mechanics (1950–1990)* (Heidelberg: Springer-Verlag, 2015), 66.

101 **"The von Neumann proof":** Charles Mann and Robert Crease, "John Bell," *Omni*, May 1988, 88.

107 **"I cannot seriously believe":** Jürgen Audretsch, *Entangled Systems: New Directions in Quantum Physics* (Weinheim, Wiley-VCH, 2007), 130.

109 **"These experiments are a magnificent affront":** Brian Greene, *The Fabric of the Cosmos: Space, Time, and the Texture of Reality* (New York: Vintage Books, 2005), 199.

110 **recognized for their efforts in 2010:** "Where Credit is Due," editorial in *Nature Physics* (Jun 1, 2010), https://www.nature.com/articles/nphys1705.

111 **was housed in Schrödinger's home:** https://www.esi.ac.at/material/Evaluation 2008.pdf.

112 **in 1913, he crossed the nearly 1,800-meter-high Maloja Pass:** Alice Calaprice, Daniel Kennefick, and Robert Schulmann, *An Einstein Encyclopedia* (Princeton: Princeton University Press, 2015), 89.

112 **nature of reality and beef cattle production:** "Physicist Designs Perfect Automotive Engine," *ScienceDaily* (Feb 27, 2003), https://www.sciencedaily.com/releases/2003/02/030227071656.htm.

112 **"The mystery is not that I'm interested":** Vimal Patel, "Cows Meet Quantum, Lifelong Learning on the Banks of the Brazos," *Texas A&M University*

Science News, November 21, 2013, http://www.science.tamu.edu/news/story
.php?story_ID=1141#.WTOOvO-0k7Y.

113 **"A dumb kid from Wyoming":** Interview of Marlan Scully by Joan Brom-
berg, July 15 and 16, 2004, Niels Bohr Library & Archives, American Institute
of Physics, College Park, MD, www.aip.org/history-programs/niels-bohr
-library/oral-histories/32147.

115 **"When the province of physical theory":** Wheeler and Zurek, *Quantum
Theory and Measurement*, 169.

115 **But by 1970, Wigner changed his mind:** Art Hobson, *Tales of the Quantum:
Understanding Physics' Most Fundamental Theory* (New York: Oxford University Press, 2017), 201.

116 **"We propose and analyze an experiment":** Marlan O. Scully and Kai Drühl,
"Quantum Eraser: A Proposed Photon Correlation Experiment Concerning
Observation and 'Delayed Choice' in Quantum Mechanics," *Physical Review
A* 25, no. 4 (Apr 1982): 2208–13.

116 **Zeilinger and colleagues:** Thomas J. Herzog et al., "Complementarity and
the Quantum Eraser," *Physical Review Letters* 75, no. 17 (Oct 23, 1995):
3034–37.

116 **Scully eventually joined hands:** Yoon-Ho Kim et al., "Delayed 'Choice'
Quantum Eraser," *Physical Review Letters* 84, no. 1 (Jan 3, 2000): 1–5.

129 **"Although we listened to hundreds":** Freire, *The Quantum Dissidents*, 20.

130 **"It looks strange":** Wheeler and Zurek, *Quantum Theory and Measurement*, 185.

130 **"The final story of the relation":** Alwyn Van der Merwe, Wojciech Hubert
Zurek, and Warner Allen Miller, eds., *Between Quantum and Cosmos: Stud-
ies and Essays in Honor of John Archibald Wheeler* (Princeton: Princeton
University Press, 2017), 10.

134 **German physicist Mauritius Renninger:** A. Cardoso, J. L. Cordovil, and
J. R. Croca, "Interaction-Free Measurements: A Complex Nonlinear Explanation," *Journal of Advanced Physics* 4, no. 3 (Sep 2015): 267–71.

138 **The paper did see the light of day:** Avshalom Elitzur and Lev Vaidman,
"Quantum Mechanical Interaction-Free Measurements," *Foundations of Physics* 23, no. 7 (Jul 1993): 987–97.

139 **"Surely . . . it can be no sin to *fail*":** Roger Penrose, *Shadows of the Mind: A
Search for the Missing Science of Consciousness* (Oxford: Oxford University
Press, 1996), 269.

143 **A nearly exact replica of this experiment:** William Irvine, Juan Hodelin,
Christoph Simon, and Dirk Bouwmeester, "Realization of Hardy's Thought
Experiment with Photons," *Physical Review Letters* 95 (Jul 15, 2005):
030401–4.

144 **His paper:** Lucien Hardy, "Quantum Mechanics, Local Realistic Theories, and Lorentz-Invariant Realistic Theories," *Physical Review Letters* 68, no. 20 (May 18, 1992): 2981–4.

144 **"simpler and more compelling":** David Mermin, "Quantum Mysteries Refined," *American Journal of Physics* 62, no. 10 (Oct 1994): 880–7.

144 **"stands in its pristine simplicity":** Ibid.

147 **"There's an entirely different way":** David Albert, *Quantum Mechanics and Experience*, 134.

147 **He founded the first school of theoretical physics:** Robert Sanders, "Conference, Exhibits Probe Science and Personality of J. Robert Oppenheimer, Father of the Atomic Bomb," *UC Berkeley News*, April 13, 2004, https://www.berkeley.edu/news/media/releases/2004/04/13_oppen.shtml.

147 **"Bohr was God and Oppie was his Prophet":** Freire, *The Quantum Dissidents*, 26.

148 **Oppenheimer reassured UC Berkeley:** Ibid.

148 **"probably Oppenheimer's best student at Berkeley":** Ibid.

148 **Princeton suspended him:** Ibid., 28.

150 **"Until we find some real evidence":** Bohm, *Quantum Theory*, 115.

150 **"only orthodox in the Copenhagen":** Karl Popper, *Quantum Theory and the Schism in Physics: From the Postscript to the Logic of Scientific Discovery* (New York: Routledge, 2013), 36.

150 **"general conceptual framework of the quantum theory":** Bohm, *Quantum Theory*, 115.

150 **"one of the most fundamental":** Ibid., 623.

150 **"no theory of . . . hidden variables":** Ibid.

156 **because it was too long:** Freire, *The Quantum Dissidents*, 31.

156 **"If I write a paper":** Ibid.

156 **"All of the objections":** David Bohm, "A Suggested Interpretation of the Quantum Theory in Terms of 'Hidden' Variables," *Physical Review* 85, no. 2 (Jan 15, 1952): 166–79.

156 **"If one man finds a diamond":** Freire, *The Quantum Dissidents*, 32.

157 **The impetus came from:** Yves Couder and Emmanuel Fort, "Single-Particle Diffraction and Interference at a Macroscopic Scale," *Physical Review Letters* 97 (Oct 13, 2006): 154101–4.

160 **"Our results do not close the door":** Giuseppe Pucci, Daniel Harris, Luiz Faria, and John Bush, "Walking Droplets Interacting with Single and Double Slits," *Journal of Fluid Mechanics* 835 (Jan 25, 2018): 1136–56.

167 **The trio published a paper showing the trajectories:** Chris Philippidis, Chris Dewdney, and Basil Hiley, "Quantum Interference and the Quantum Potential," *Il Nuovo Cimento B* 52, no. 1 (Jul 1979): 15–28.

169 **In 1988, Yakir Aharonov:** Yakir Aharonov, David Albert, and Lev Vaidman, "How the Result of a Measurement of a Component of the Spin of a Spin-1/2 Particle Can Turn Out to Be 100," *Physical Review Letters* 60 (Apr 4, 1988): 1351–54.

170 **"It must be emphasized":** Howard Wiseman, "Grounding Bohmian Mechanics in Weak Values and Bayesianism," *New Journal of Physics* 9 (Jun 2007): 165.

173 **"The team is the first to track":** Hamish Johnston, "Physics World Reveals Its Top 10 Breakthroughs for 2011," *Physics World* (Dec 16, 2011), http://physics world.com/cws/article/news/2011/dec/16/physics-world-reveals-its-top -10-breakthroughs-for-2011.

175 **"Tersely: Bohm trajectories are not realistic":** Berthold-Georg Englert, Marlan Scully, Georg Süssmann, and Herbert Walther, "Surrealistic Bohm Trajectories," *Zeitschrift für Naturforschung A* 47, no. 12 (1992), 1175–86.

177 **It required a small but significant:** Dylan Mahler et al., "Experimental Non-local and Surreal Bohmian Trajectories," *Science Advances* 2, no. 2 (Feb 19, 2016): e1501466.

183 **"inspired by the modern realization":** Louis Sass, *Madness and Modernism: Insanity in the Light of Modern Art, Literature, and Thought* (Cambridge, MA: Harvard University Press, 1994), 31.

187 **"A university student attending lectures":** Carlo Rovelli, *Reality Is Not What It Seems: The Journey to Quantum Gravity* (New York: Riverhead Books, 2017), 148.

189 **Penrose imagines a photon:** Roger Penrose, *Fashion, Faith, and Fantasy in the New Physics of the Universe* (Princeton: Princeton University Press, 2016), 162.

193 **"shifty split":** John Bell, "Against 'Measurement,'" *Physics World* 3, no. 8 (Aug 1990): 33.

193 **the Hungarian physicist:** Lajos Diósi, "A Universal Master Equation for the Gravitational Violation of Quantum Mechanics," *Physics Letters A* 120, no. 8 (Mar 16, 1987): 377–81.

193 **three physicists:** Giancarlo Ghirardi, Alberto Rimini, and Tullio Weber, "Unified Dynamics for Microscopic and Macroscopic Systems," *Physical Review D* 34, no. 2 (Jul 15, 1986): 470–91.

195 **"Any embarrassing macroscopic ambiguity":** John Bell, *Speakable and Unspeakable in Quantum Mechanics* (Cambridge: Cambridge University Press, 1989), 204.

195 **"have a certain kind of goodness":** John Bell quoted in Giancarlo Ghirardi, *Sneaking a Look at God's Cards: Unraveling the Mysteries of Quantum Mechanics* (Princeton: Princeton University Press, 2005), 415.

195 **"They are honest attempts":** Ibid.

197 **In 1991, Jürgen Mlynek and colleagues:** O. Carnal and J. Mlynek, "Young's Double-Slit Experiment with Atoms: A Simple Atom Interferometer," *Physical Review Letters* 66, no. 21 (May 27, 1991): 2689.

197 **atoms could be diffracted at gratings:** Philip Moskowitz, Phillip Gould, Susan Atlas, and David Pritchard, "Diffraction of an Atomic Beam by Standing-Wave Radiation," *Physical Review Letters* 51, no. 5 (Aug 1, 1983): 370.

197 **a double-slit experiment done with neon atoms:** Fujio Shimizu, Kazuko Shimizu, and Hiroshi Takuma, "Double-Slit Interference with Ultracold Metastable Neon Atoms," *Physical Review A* 46, no. 1 (Jul 1, 1992): R17.

198 **In 1999, Zeilinger, Arndt, and their team:** Markus Arndt et al., "Wave–Particle Duality of C_{60} Molecules," *Nature* 401 (Oct 14, 1999): 680–82.

199 **it's a bespoke molecule:** Sandra Eibenberger et al., "Matter–Wave Interference of Particles Selected from a Molecular Library with Masses Exceeding 10,000 amu," *Physical Chemistry Chemical Physics* 15 (Jul 8, 2013): 14696-700.

204 **There's a 146-meter-high tower in Bremen:** https://www.zarm.uni-bremen.de/en/drop-tower/general-information.html.

207 **Penrose had plans:** Roger Penrose, "Wavefunction Collapse as a Real Gravitational Effect," in *Mathematical Physics 2000*, ed. A. Fokas, A. Grigoryan, T. Kibble, and B. Zegarlinski (London: Imperial College Press, 2000), 266 82.

210 **He knew people at NASA:** https://youtube/mvHg5PcXb6k?t=45.

213 **When they wrote their paper:** William Marshall, Christoph Simon, Roger Penrose, and Dirk Bouwmeester, "Towards Quantum Superpositions of a Mirror," *Physical Review Letters* 91, no. 13 (Sep 23, 2003): 130401.

216 **"The collapse [of the wavefunction]":** Simon Saunders, Jonathan Barrett, Adrian Kent, and David Wallace, eds., *Many Worlds?: Everett, Quantum Theory, and Reality* (Oxford: Oxford University Press, 2010), 582.

216 **"He was rather upset when I met him":** Dirk Bouwmeester speaking at the Institute for Quantum Computing in Waterloo, Ontario, Canada, May 2013, https://youtube/g7RqLbqDr4U?t=387.

217 **"Actualities seem to float":** William James, *The Will to Believe: And Other Essays in Popular Philosophy* (New York: Longmans Green and Co, 1907), 151.

218 **Wheeler's attitude likely rubbed:** Transcript of conversation between Hugh Everett and Charles Misner, in Hugh Everett III, *The Everett Interpretation of Quantum Mechanics: Collected Works 1955–1980 with Commentary*, ed. Jeffrey A. Barrett and Peter Byrne (Princeton: Princeton University Press, 2012), 309.

219 **"What actually does happen":** Everett, *The Everett Interpretation*, 65.

220 **"In other words, the observer":** Ibid., 67.

221 **"for almost all of the":** Ibid., 69.

222 **"can lay claim to a certain"**: Ibid.

222 **"One can imagine an intelligent"**: Ibid., 69–70.

222 **"I am frankly bashful"**: Ibid., 71.

223 **"We do not believe"**: Ibid., 153.

223 **"objectionable" dualism**: Ibid.

223 **An American physicist, Alexander Stern**: Ibid., 214.

223 **"lack meaningful content"**: Ibid., 215.

223 **"matter of theology"**: Ibid., 217.

223 **"I would not have imposed"**: Ibid., 219.

224 **"this very fine and able"**: Ibid.

224 **"javelin proof"**: Ibid., 212.

224 **"I was stunned, I was shocked"**: Interview of Bryce DeWitt and Cecile DeWitt-Morette by Kenneth W. Ford, February 28, 1995, Niels Bohr Library & Archives, American Institute of Physics, Oral Histories, https://www.aip .org/history-programs/niels-bohr-library/oral-histories/23199.

224 **"I can testify to this"**: Everett, *The Everett Interpretation*, 246.

225 **"hopelessly incomplete"**: Ibid., 255.

225 **"a philosophic monstrosity"**: Ibid.

225 **"From the viewpoint of the theory"**: Ibid., 254.

225 **a conference in October 1962 in Cincinnati, Ohio**: A transcript of the conference is available in Everett, *The Everett Interpretation*, 270.

225 **"It seems to me that if this is the case"**: Everett, *The Everett Interpretation*, 273.

226 **"Somehow or other we have here"**: Ibid.

226 **"Yes, it's a consequence"**: Ibid., 274.

226 **"You eliminate one of the two"**: Ibid., 275.

226 **"Each individual branch"**: Ibid, 276.

226 **"This universe is constantly splitting"**: Bryce DeWitt, "Quantum Mechanics and Reality," *Physics Today* 23, no. 9 (Sep 1970): 30.

227 **"I still recall vividly the shock"**: Ibid.

227 **Universe Splitter**: https://itunes.apple.com/us/app/universe-splitter/id3292 33299.

234 **"started to sound a little bit too practical"**: David Wallace, "The Emergent Multiverse: The Plurality of Worlds—Quantum Mechanics," February 21, 2015, https://youtube/2OoRdyn2M9A?t=183.

236 **"I think one should invoke"**: Everett, *The Everett Interpretation*, 278.

236 **"So the parallel universes are cheap"**: Paul Davies and Julian Brown, eds., *The Ghost in the Atom* (Cambridge: Cambridge University Press, 1993), 84.

238 **"if the other universes"**: Frank Wilczek, "Remarks on Energy in the Many Worlds," Center for Theoretical Physics, MIT, Cambridge, Massachusetts, July 24, 2013, http://frankwilczek.com/2013/multiverseEnergy01.pdf.

240 **probability as something subjective:** Charles Sebens and Sean Carroll, "Self-Locating Uncertainty and the Origin of Probability in Everettian Quantum Mechanics," *British Journal for the Philosophy of Science* 69, no. 1 (Mar 1, 2018): 25–74.

243 **"rather than relinquishing":** Christopher A. Fuchs, "On Participatory Realism," June 28, 2016, https://arxiv.org/abs/1601.04360.

244 **"built on billions upon billions":** John Wheeler paraphrased in Ibid.

245 **Quantum Bayesianism:** Carlton Caves, Christopher Fuchs, and Rüdiger Schack, "Quantum Probabilities as Bayesian Probabilities," *Physical Review A* 65, no. 2 (Jan 4, 2002): 022305.

246 **Einstein is thought to have:** Matthew Leifer, "Is the Quantum State Real? An Extended Review of ψ-ontology Theorems," *Quanta* 3, no. 1 (Nov 2014): 72.

249 **note that the same argument:** Ibid.

251 **"from the third-person perspective":** David Wallace, *The Emergent Multiverse: Quantum Theory according to the Everett Interpretation* (Oxford: Oxford University Press, 2012), 310.

251 **"From the perspective of a given experimenter":** Ibid.

252 **"Quantum mechanics, in the QBist interpretation":** Christopher Fuchs, David Mermin, and Rüdiger Schack, "An Introduction to QBism with an Application to the Locality of Quantum Mechanics," November 20, 2013, https://arxiv.org/pdf/1311.5253.pdf.

252 **"Science is about the interface":** David Mermin, "Why QBism Is Not the Copenhagen Interpretation and What John Bell Might Have Thought of It," September 8, 2014, https://arxiv.org/pdf/1409.2454.pdf.

253 **"as a nutcase":** Christopher Fuchs, *Coming of Age with Quantum Information: Notes on a Paulian Idea* (Cambridge: Cambridge University Press, 2011), Kindle edition.

253 **"We need to talk":** Ibid.

253 **"Great talk!":** Ibid.

253 **"No, it wasn't!":** Ibid.

253 **"You know what I would do":** Ibid.

255 **"The est foundation's physics conferences":** David Kaiser, *How the Hippies Saved Physics: Science, Counterculture, and the Quantum Revival* (New York: Norton, 2011), 189.

256 **In one talk, he asked his audience to consider two theories:** Richard Feynman Messenger Lectures on the Character of Physical Law, Lecture 7, "Seeking New Laws," November 1964, http://www.cornell.edu/video/richard-feynman-messenger-lecture-7-seeking-new-laws.

256 **"In order to get new theories":** Ibid.

256 **"In other words, although they are identical":** Ibid.

257 **This inspired Wiseman:** Jay Gambetta and Howard Wiseman, "Interpretation of Non-Markovian Stochastic Schrödinger Equations as a Hidden Variable Theory," *Physical Review A* 68 (Dec 9, 2003): 062104.

259 **The results they obtained are eerily similar:** Michael Hall, Dirk-André Deckert, and Howard Wiseman, "Quantum Phenomena Modeled by Interactions between Many Classical Worlds," *Physical Review X* 4 (Oct 23, 2014): 041013.

263 **"If Born's rule fails":** W. H. Zurek, quoted in Urbasi Sinha et al., "A Triple Slit Test for Quantum Mechanics," *Physics in Canada* 66, no. 2 (Apr/Jun 2010): 83.

263 **sometimes the triple-slit:** G. Rengaraj et al., "Measuring the Deviation from the Superposition Principle in Interference Experiments," November 20, 2017, https://arxiv.org/abs/1610.09143.

264 **Or is it equal to the amplitude:** Rahul Sawant et al., "Nonclassical Paths in Quantum Interference Experiments," *Physical Review Letters* 113, no. 12 (Sep 19, 2014): 120406.

266 **"Any . . . situation in quantum mechanics":** Feynman Messenger Lectures, Lecture 6, "Probability and Uncertainty: The Quantum Mechanical View of Nature," http://www.cornell.edu/video/richard-feynman-messenger-lecture -6-probability-uncertainty-quantum-mechanical-view-nature.

ACKNOWLEDGMENTS

I remember being thrilled by Gary Zukav's *Dancing Wu Li Masters* when I read it in the 1980s. The mysteries of quantum physics came alive. The book had, of course, a description of the double-slit experiment, besides a lot else. Then, as a journalist, I too started writing stories about quantum mechanics and encountered the iconic experiment at every turn. An idea took shape: a story about quantum physics told from the perspective of the double-slit experiment. But it remained on a back burner for years.

Thanks to my editor, Stephen Morrow, for seeing the possibilities and making me revisit the idea and then seeing the book through to the end. Thanks also to Madeline Newquist and others at Dutton and to my agent, Peter Tallack, for their help in making this happen.

A book like this needs an illustrator. Thanks to my friend Ajai Narendran for introducing me to Roshan Shakeel. Roshan turned my half-baked sketches into clean, sharp drawings that wonderfully complement the words. I'm grateful to Roshan for his unflagging efforts, and to Ajai for supporting and encouraging both me and Roshan.

I'm grateful also to Rob Sunderland, Lis Rasmussen, Felicity Pors, and the rest of the staff at the Niels Bohr Archive in Copenhagen, Denmark, for all their help in accessing historical documents.

ACKNOWLEDGMENTS

I leaned heavily on physicists to explain quantum mechanics to me. They took the time, either in person or by phone or email, to enlighten me on the many conceptual issues that make quantum physics so confounding and enthralling; many read parts of the book and caught errors and suggested changes. I'm grateful to (in order of the chapters): Lucien Hardy, Alain Aspect, Philippe Grangier, David Albert, Tim Maudlin, Anton Zeilinger, Marlan Scully, Rupert Ursin, Xiao-Song Ma, Lev Vaidman, Sheldon Goldstein, John Bush, Tomas Bohr, Chris Dewdney, Basil Hiley, Aephraim Steinberg (for many discussions and meetings over the years), Roderich Tumulka, Roger Penrose, Markus Arndt, Dirk Bouwmeester, Sean Carroll, David Wallace, Howard Wiseman, Chris Fuchs, David Mermin, David Kaiser, Leonard Susskind, Urbasi Sinha, John Sipe, and Neal Abraham.

I'm especially grateful to John Bush for vetting almost all the chapters. His enthusiasm was infectious. And a special thanks to Antoine Tilloy for reading and commenting on the whole book. Thanks also to my friends Sriram Srinivasan and Varun Bhatta for their input. Most of all, thanks to Adam Becker for his support throughout the writing of this book—including animated discussions over innumerable coffees and lunches in Berkeley—and for catching some errors in the final draft.

Any errors that remain are, of course, my responsibility.

Thanks to Banu and Ramesh for hosting me for months in Arlington and Amherst, as I went about meeting the coterie of quantum physicists on the US East Coast; to Caroline Sidi for hosting me in Paris; to Gita Suchak for making me feel at home in London; and to Rao Akella for help finding quotes for the epigraphs. And last, but by no means least, my thanks to my family back in India for their support, particularly my parents.

INDEX

INDEX

ABOUT THE AUTHOR

Anil Ananthaswamy is an award-winning journalist and former staff writer and deputy news editor for the London-based *New Scientist* magazine. He has been a guest editor for the science writing program at the University of California, Santa Cruz, and organizes and teaches an annual science journalism workshop at the National Centre for Biological Sciences in Bengaluru, India. He is a freelance feature editor for the Proceedings of the National Academy of Science's *Front Matter.* He contributes regularly to *New Scientist,* and has also written for *Nature, National Geographic News, Discover, Nautilus, Matter, The Wall Street Journal,* and the UK's *Literary Review.* His first book, *The Edge of Physics,* was voted book of the year in 2010 by *Physics World,* and his second book, *The Man Who Wasn't There,* won a Nautilus Book Award in 2015 and was longlisted for the 2016 PEN/E. O. Wilson Literary Science Writing Award.